South Asian Technospaces

Digital Formations

Steve Jones
General Editor

Vol. 36

PETER LANG
New York • Washington, D.C./Baltimore • Bern
Frankfurt am Main • Berlin • Brussels • Vienna • Oxford

South Asian Technospaces

Edited by Radhika Gajjala
& Venkataramana Gajjala
WITH Natalia Rybas

PETER LANG
New York • Washington, D.C./Baltimore • Bern
Frankfurt am Main • Berlin • Brussels • Vienna • Oxford

Library of Congress Cataloging-in-Publication Data

South Asian technospaces / edited by Radhika Gajjala,
Venkataramana Gajjala.
p. cm. — (Digital Formations; v. 36)
Includes bibliographical references and index.
1. Information society—South Asia. 2. Internet users—South Asia.
3. Digital divide—South Asia. I. Gajjala, Radhika.
II. Gajjala, Venkataramana.
HM851.S68 303.48'330954—dc22 2008003854
ISBN 978-1-4331-0244-8 (hardcover)
ISBN 978-0-8204-8122-7 (paperback)
ISSN 1526-3169

Bibliographic information published by **Die Deutsche Bibliothek**.
Die Deutsche Bibliothek lists this publication in the "Deutsche
Nationalbibliografie"; detailed bibliographic data is available
on the Internet at http://dnb.ddb.de/.

Cover design by Pratap Gajjala

The paper in this book meets the guidelines for permanence and durability
of the Committee on Production Guidelines for Book Longevity
of the Council of Library Resources.

29 Broadway, 18th floor, New York, NY 10006
www.peterlang.com

Printed in the United States of America

This project is dedicated to our Fathers
(Vedula Narasimha Murthy and Gajjala Jagathpathi),
whose careers led us to our traveling lives and whose
shared poetry carved the path for our journey together.
—VENKATARAMANA GAJJALA & RADHIKA GAJJALA

For Sergey
—NATALIA RYBAS

CONTENTS

PREFACE

For the past decade, I have smiled when a message from "cyberdiva.org," also known as Radhika Gajjala, appears in my email inbox. Through collaborative research, publications, panels, workshops, talks and events—not to mention annual forays into whatever our students find to be the new "killer app" in order to inspire inter-institutional digital production that critically engages the relationship between ICT and race, gender, sexuality, space and place—I've been the beneficiary of Gajjala's energy and wisdom. I know well that Gajjala, in cahoots with her cyberdiva avatar, is passionately committed to trouble: asking the questions that are easier left unsaid, troubling the assumptions both she and others take for granted. In compiling *South Asian Technospaces*, Radhika teams up with her longtime co-conspirator, Vinnie Gajjala, to trouble what we understand as digital divides—no longer thinking of this as a problem that can be discussed solely in terms of hardware and software access, knowledge and practical application, margin and center. The editors challenge the meaning of categories such as postcolonial, South Asian, American, homeland, diaspora, feminism, scholarly thought, and technological literacy, while their questions dwell, even if uneasily so, in the the chasms between theory and practice, center and periphery, and academic work and play. By offering this nuanced and multifaceted treatment of technological communication, they challenge their readers to imagine disciplinary gaps not as unresolvable problems but as bountiful opportunities for thinking about the world at this particular historical juncture.

By and large, critical discussion of globalization focuses on economic markets and their proliferation, thus it comes as no surprise that writings from the disciplines of economics, geography, political science, history, philosophy and critical theory are frequently suffused with despair. Of course, the inequities inherent in the spread of Western dominance of the global marketplace, the increasing importance placed upon visual displays of Euro-American style accumulation, the persistence of nationalism and xenophobia, and the ever increasing gap that separates the rich from the poor provide a trenchant base for overall skepticism. The tenor of utopianism found in the interdisciplinary field of cyberculture studies/new media studies stands in stark contrast to these other fields of inquiry. Here, scholarship indefatigably presumes a mutually beneficial link between technology (read: information and communications technologies, ICT) and globalization. In other words, more communication is inherently good: more people, more access, more data exchange. Well, from a liberal, capitalist vantage point predicated upon the sanctity of individual liberties as well as the right of consumers to stockpile wealth and goods, this can only be interpreted as a boon. Increasing access and widespread participation in ICT will bridge the "digital divide," allow a cacophony of voices to participate in cyber-community building, encourage intercultural awareness, put an end to racisms and cultural domination throughout the world, and abolish fear of the other. Or so the story goes.

Although it would seem that simply deploying methodologies taken from the social sciences and cultural studies to explore media-related concerns would render this incommensurability visible, it rarely does. Despite the interdisciplinarity of cyberculture studies and its inextricable interconnectedness with these other disciplines, the schism in their ideological trajectories—globalization as exploitative versus technology as an answer to global inequities—has led to a dearth of scholarship that takes as its foundation the specific instances of ICT use on extremely local groups and the implications this has on our understanding of the contours of globalization. The vicissitudes of ICT use, the how, when, and why of particular forms of access, the varying conditions of embodiment vis-à-vis digital interfaces, accounts of the changing nature of digital communication, and the ways in which digital communication challenges contemporary theoretical knowledge: these are areas that this book sets out to explore.

This collection of essays provides an important contribution to cyberculture studies. Under the rubric of "South Asian Diaspora," it isolates the processes of digital communications technologies and their relations to globalization. While the scope of the collection may seem limited—digital practices among an assortment of South Asian-identifying populations—its implications and showcasing of methodological approaches speak to all those concerned with the theories and practices of digital communication. By assembling contributions that consider technology from a variety of angles, Gajjala and Gajjala reveal the blind spots of new media theories, postcolonial theory, and diaspora studies. They weave these varied contributions together, crafting a cohesive collection that demonstrates the remarkable woof and warp of labor, education, and communication strategies that mark the advent of worldwide participation in digital technologies.

South Asian Technospaces highlights the disjunctures of new media practices, troubling even the most basic categories that are utilized in many of the book's essays. In the process, the individual contributions provide complex glimpses of how digital technologies function in particular locales. From the Dalits of India to the players of Silicon Valley, place and space, "real" and "virtual," online and offline are revealed as artificial constructions that no longer are usefully conceived of as oppositional. The authors also contest hegemonic conceptualizations of diaspora—unearthing their limitations by showing diasporic conditions faced by non-dispersed populations whose labor is outsourced, for instance, or Bangladeshi and Pakistani women in East London whose dissimilar circumstances lead to uneven experiences of technology-centered education. This suggests to readers that our contemporary moment requires a constant renegotiation of even our most common nomenclature for traveling cultures and laboring bodies. Accordingly, the pages that follow open up the multifaceted layers that emerge at the interstices of virtual communities and South Asian diasporas. Diaspora, dispersal and immersion in new cultures should no longer be read as singular processes following a particular set of physical, psychic, and material avenues.

At times, the essays depict the potential for South Asian participation as almost celebratory, revealing strategic responses to a market eager to find low cost, high skilled workers, but dominated by multi-national corporations with little knowledge of cultural conventions beyond the boardroom walls. Other essays remind us of the continuing imbalances produced even, and especially in the aftermath of decolonization, teasing out the ever-persistent and increasingly pernicious effects of capitalism's manifold ideologies. Spending all evening at a computer terminal in a call center in Bangalore city may not make life better when one must embrace U.S. culture by night and remain seamlessly bounded to local culture by day. The access to digital technologies and remuneration for expertise within systems and culture are highly differentiated.

As a collection, this work takes a broad approach to ICT by including Internet use, software development, telephony, music and film. The framing of these essays allows us to consider convergence in a multitude of manifestations, moving us from the particular in order to gesture towards the sum, imagining both individuals and vast theoretical categories in better relation. *South Asian Technospaces*, in fact, meets Gajjala's concluding challenge: that Internet and cyberculture researchers must "re-examine conceptual categories and frameworks such as 'diaspora,' 'globalization,' 'new media' and even 'empowerment,' 'multiculturalism' and 'voice/voicelessness.'" The anthology deftly problematizes the placement of bodies in dispersal—showing other diasporic conditions that, rather than talking generally about the machinations of power, leads us to a better understanding of the multifarious ways in which technologically mediated practices operate in specific locales. Illuminating such a specialized topic, South Asian technospaces, from a strategically placed range of viewpoints shows us nothing less than how the focus on the particular precipitates and undergirds any understanding of the machinations and configurations that define the complex flows of global digital communication.

Jillana Enteen
Evanston, November 2007

INTRODUCTION

South Asian Digital Networks and Global Technospaces

Radhika Gajjala and Venkataramana Gajjala

MOST ECONOMIC and social activity in our daily lives somehow leads us back to the Internet: clicking on links, "going" to Web sites, "meeting" friends online, "chatting" with a sister or brother, shopping, sending greetings for Christmas, Diwali, or Id, "going" to work while hanging out with family during Thanksgiving break, SMSing a colleague while running to get the children to school on time....Metaphors of space and movement in relation to networked technologies have become commonplace as we have come to take our digitally mediated techno-habitus for granted. This edited collection on South Asian technospaces provides perspectives on how South Asian (and often more specifically Indian) diasporas inhabit technomediated environments through their economic and sociocultural activities. The themes examined include religion, caste, language, and gender in online communities; call centers and their role in the global economy; Bollywood online and offline; digital music; and Web sites to help arrange marriages. This book maps "South Asia" in relation to global technospaces produced through and as a consequence of economic globalization.

The label "South Asian" has been problematized by postcolonial studies scholars, and we too use it strategically and with caution, even as some contributors use the term "Indian diaspora" to refer to the post-1960s migration from the nation called India. While noting the complexity of talking about South Asian diasporas, we also wish to highlight how South Asian and Indian diasporas are shaped by and shape Asian, African, and Southeast Asian diasporas. Linda Leung's and Nabeel Zuberi's chapters in this volume are critical in this respect insofar as they bring all these intersections into the dialogue about South Asian and Indian technospaces in relation to globalization. They show how these labels—Asian, African, black, Indian, Pakistani, Bangladeshi, South Asian—mesh and intertwine in global technomediated contexts. Globalization processes—material and discursive hegemonies produced at the intersection of the economic, the cultural, and the social—feed into economic and cultural local formations

The use of "diaspora" in relation to online networks is probably more problematic still. However, just as Internet researchers have taken to the phrase "virtual community" and use it with ease (no longer feeling the need to critique or problematize it), several of us who research online networks of South Asians and Indians tend to use the term "diaspora" quite easily. Rather than offer definitions at length in this introductory chapter, the co-editors have written chapters situating the notion of diasporic networks and historicizing the notion of (Indian) diaspora in relation to the economic context (Chapter 2) and examining the concept of virtual community and imagined community in relation to Indian/South Asian diasporas (Chapter 3).

These two chapters by the editors follow Leung's, which discusses "Asian" as a label that is used in relation to immigrants to England from the South Asian subcontinent. Leung's chapter occupies class, gender, and race intersections while examining ethnic minorities in England from what we might refer to as Indian, Pakistani, and Bangladeshi diasporas who self-identified as Asian. Chapter 4, by Zuberi, refers to South Asian diasporas to African nations and discusses how black identities are interwoven by talking about the dialogics of South Asian and African diasporas. Thus he notes a "large number of hip-hop, R&B, and Jamaican dancehall tracks that integrate South Asian instrumentation, refrains, and voices, primarily from bhangra and Hindi film song."

When we put out a call for chapters, we hoped to make this a collection that truly engaged South Asian diasporas. In fact, a majority of the essays we received were specifically about Indian contexts. Therefore, the collection gives us more of a picture of Indian technospaces in general. However, the chapters by Leung and Zuberi provide us with an understanding of the complexity of South Asian diasporas old and new by pointing to how Asian and African identities have historically connected with South Asian digital diasporas.

We felt that the inclusion of two chapters that would not fit neatly into such a clear (yet false) picture of Indian diasporas would serve to problematize the notion of an Indian diaspora. To articulate South Asian diasporas as "Indian" diasporas, "Pakistani" diasporas, "Sri Lankan" diasporas, "Bangladeshi" diasporas, and so on would shift the lens and direct our focus in ways that might or might not be productive in an effort to understand how peoples from these locations traveled before postcolonial national formations. Furthermore, some methods of articulating the specificities of diasporic routes might lead us to explore regional, linguistic, or even caste-based labels rather than national ones in referring to diasporas. Thus we might begin to look at terms such as "Tamil diasporas" or "Dalit diasporas" or "Punjabi diasporas" or "Telugu diasporas," and once again, even in this kind of sorting, "Bangladeshi diasporas." Of course, Bangladeshi diasporas and Tamil diasporas, understood as linguistic rather than geographic or religious categories, would cross national borders. While a Bangladeshi diaspora could refer to the nation of Bangladesh, it could also mean the speakers of Bengali (Bangla) and include people from Bengal, India. Furthermore, while a Tamil diaspora has been most often associated with Sri Lankan Tamils and an "Eelam diaspora" (Enteen, 2006; Jeganathan, 1998), the Tamil diaspora also includes Indians from the state of Tamil Nadu.

Zuberi's chapter, which examines configurations of audio space and time in digital recordings as they articulate black Atlantic and South Asian elements in different ways, is followed by Amit Rai's chapter on the "purple pleasures" of Bollywood, in which he examines

> the folding of cinema into various computer interfaces and across technological platforms, which has both necessitated and facilitated the globalization of film finance as well as changes in censorship legislation and practice from "censorship" to "certification."

Critical Internet research and cyberculture studies overlap and draw from a variety of disciplinary intersections. Therefore this volume is not organized along disciplinary divisions or along lines of Internet research, cyberculture studies, cultural studies, postcolonial studies, and so on. The collection is interdisciplinary because the themes taken up by the authors are at various online/offline and local/global intersections.

Chapter 6 brings in the Dalit question. P. Thirumal uses the case of an Internet campaign to show how issues related to Dalits' oppression and struggles with institutional power weave in and out of global and national contexts while impacting the local material context in complex ways. He poses interrelated questions about technology, development, and democracy in relation to the public controversy surrounding the alleged beating of a Dalit employee in a reputed NGO in India during 2003.

In Chapter 7, Vinita Agarwal and Patrice Buzzanell talk about spatialization and marginalization. The chapter is focused on young girls in a *basti* whose coming-of-age experiences

> seem characterized by their multiple and contradictory status of observers, participants, and maybe even victims of the push and pull of the larger social forces of chaos and order, control and vulnerability, death and hope, [but] are by no means unique to the specific geographical and temporal context which they describe. The duality of modernistic binaries may well characterize the forces defining marginalization and displacement in contexts of urbanization—indeed, to some extent, in all contexts of reterritorialization.

Agarwal and Buzzanell focus on reterritorialization within the Indian nation and examine Othered identities through "discursive representations of the windows and the lanes in the journals, lying at the edge of a technologically defined space (i.e., *cybermohalla*) and creating an imaginary geography of narratives."

Archana Sharma's chapter, which examines matrimonial Web sites, looks at how marriage-arranging traditions are reproduced and reshaped at the intersection of East and West in online spaces where embodied/disembodied negotiations of cultural practices serve to show how gender, nation, caste, and diaspora are reproduced in particular hegemonies.

Developing some themes implied in Sharma's chapter, Rohit Chopra examines online Hindu fundamentalist discourses. His chapter considers "the implications of the particularities of online Hindu nationalism for Hindu nationalist discourse at large and a reflection on the role of the Internet in the articulation of the global Indian-Hindu nation." In Chapter 10, Anustup Nayak and Natalia Rybas also examine a community formed online—an online transnational community formed around the state of Orissa in India.

The following three chapters examine business process outsourcing from various angles. Ananda Mitra's chapter continues the theme of outsourcing in relation to South Asian technospaces. He draws a parallel between the historic diasporic experience and the diasporic experience facilitated by the phenomenal growth in outsourcing. Whereas the former involves the physical movement of people, the latter involves the movement of process. Mitra looks specifically at the process of outsourcing call center work to India and finds that although workers in globalized call centers do not physically "travel" to a new "real" place, their job description makes it imperative that they adapt themselves to a new culture when they are in their workspace—that is, they have to "talk the talk" and "walk the walk" of their clients. He argues that technology—specifically the Internet—has blurred the distinction between the historic or traditional diasporic experience and the call center experience, because the latter, in addition to being an economic issue, is as much a political, social, and cultural issue as the former. Call center employees are exposed to the same struggles (of race, oppression, Otherness, etc.) that diasporas traditionally face, despite the absence of movement from one place to another.

Kiran Mirchandani looks at the practices of globalization and finds in her review of the literature that "understanding global processes has traditionally been limited to analyses of cross-border processes such as international trade and investment." She notes that this "dominant" paradigm involves a "top-down" process of globalization or "transnational centrism" and presupposes the immense power of the multinational corporation leading to a globalization of markets worldwide. But, as Mirchandani states, theorists have questioned this view and have emphasized the need to concentrate instead on the practices—the systems, norms, and work relations that shape transnational workers' experiences—through which what is known as "globalization" is constantly being created. She studies these practices in the call center industry in India and finds that transnational corporate practices in globalized call centers are continuously under formation.

In the final chapter, Divya McMillin examines the neocolonial processes that result in the cultural transformation of "the average Indian worker into the ideal service provider in the global marketplace." The broad questions that she poses are what shapes this transformation and what inferences can one draw from this process about the future? She conducted participant-observation and in-depth interviews at call centers in Bangalore, India, and found that Tom Freidman's belief that outsourcing to developing countries in some way results in the worker becoming independent is erroneous. She argues that (temporary) economic gains are not sufficient and that call centers afford sufficient proof of how the natives are coached by multinational corporations to serve them, continuing the age-old domination by the West. Through such analyses, we understand that while the world is not flat, the complexities and nuances of multinational corporations and outsourced work reconfigure hegemonies to make it seem flat within certain frameworks.

Note

The chapters in this volume can be used individually or in different combinations for courses on globalization and outsourcing, South Asian diasporas, Internet research and South Asia, intercultural communication, or postcolonial theory.

CHAPTER ONE

From "Victims of the Digital Divide" to "Techno-Elites": Gender, Class, and Contested "Asianness" in Online and Offline Geographies

Linda Leung

IT IS WELL DOCUMENTED that cyberspace is not as white as it was once thought to be, with evidence that the number of Internet users in Asia is greater than in the rest of the world (ClickZ Stats, 2006b; Miniwatts Marketing Group, 2006). Therefore, "Asianness" arguably has a prominent profile in cyberspace. However, one of the main limitations of the study of Asian online identity and activity is that it has been confined to a narrow socioeconomic demographic. It is often restricted to those who are advantaged in their capacity to be economic migrants: those who study overseas and remain in the countries (mainly the United States) in which they were educated, working in the professions for which they have been highly trained (Gajjala, 1999; Mallapragada, 2000; Melkote & Liu, 2000; Mitra, 1997). While the profile of cyberspace is not as white as was once claimed, the focus is still squarely on the West, the affluent, and the educated.

Within Asia itself, the focus has been on online populations in the offline north, particularly China and Japan, as these nations tend to be regarded as the most significant in terms of online participation because they represent the largest consumer markets (Greenspan, 2002). Only one South Asian country, India, is represented in the top fifteen online populations (ClickZ Stats, 2006a).

This article focuses on those who are invisible at this level of the Asian "techno-elite," also known as "victims of the digital divide." What happens when information and communication technologies (ICTs) are provided to those who would otherwise not have access to them? The particular case studied in this chapter is of two working-class ethnic minority women residing in Britain, Rosie and Noori[1], both mothers with child care responsibilities and minimal experience of computing who did not own their own computers. Asianness in a British context specifically refers to South Asian ethnicity, that is, to people originating from India, Pakistan, and Bangladesh (HMSO, 1991). In this sense, their minority status was magnified by being outside both the United States and Asia, and by being from South Asia when most scrutiny has been focused on the Asian north.

Rosie's and Noori's negotiation of the anomalies between their online and offline contexts is examined, including their migration from their countries of birth (Pakistan and Bangladesh, respectively) compared with the difficulty of escaping their socioeconomic status in Britain. Their class immobility is juxtaposed with their virtual movements through the digital diaspora. Despite their location within the ethnically dense and diverse population of East London in which Asian (i.e., Indian, Pakistani, and Banglasdeshi) ethnicities are more prevalent than black ethnicities (HMSO, 1991; Khan, 1988), they found themselves a minority (again) within the wider group of women studying new technology, the majority of whom categorized themselves as of black, African, or Caribbean origin.

In the context of postmodern notions of fluid, transformational identities in virtual environments and the contested definitions of British Asianness, the chapter explores how Rosie and Noori find some anchorage in the politics of identity through their forays into cyberspace. How do they make connections with others in their consumption of online texts? What does the online world offer, and what kind of technological literacies must they develop before they can successfully participate in it?

Drawing from interview material, written autobiographical responses, and textual analyses, the chapter also looks at how Rosie's and Noori's ethnicities are "fixed" or "grounded" by their relationships to the other members of the group.

The Places and (Cyber)spaces of Asia Online

In examining Asianness in the online world, it is somewhat ironic that the body of quantitative and qualitative research data about Internet participation focuses almost exclusively on the United States. It is well known that Asian Americans' Internet participation is higher than that of any other ethnic group in the United States (ClickZ Stats, 2001). While this participation is somewhat male dominated, the proportion of Asian American women online is nonetheless higher than that of any other ethnic group. Similarly, although Asian Americans have a young online profile, they are the most experienced Internet users. Their socio-economic profile and specific daily Internet activities have been studied in detail. Arguably, more is known about Asian American Internet audiences than about any other type of Asian online participation. Such research tends to contextualize Asian Americans among three main ethnic groups that have an online presence, the others being Hispanic and African American Internet users. However, this categorization immediately renders invisible and problematic the spectrum of Asian ethnic minority groups in the United States alone participating in cyberspace. Also, the U.S.-centricity of these studies seems somewhat misplaced if one compares the size of the Asian American online constituency relative to the U.S. Internet population with the respective size of the Asian online population (Miniwatts Marketing Group, 2006). Internet usage in Asia easily surpasses that of the rest of the world (Miniwatts Marketing Group, 2006), yet little is known about the online behavior of users from Asian countries.

Evidently, the infrastructure, processes, and practices necessary to assess online participation in Asia are not as established as in the United States, which has a number of sources of data (such as the Pew Internet & American Life Project, the CIA's World Factbook, and Nielsen/Netratings). What this means is that despite the pervasive online presence in Asia, attention still resides overwhelmingly with American (specifically U.S.) Asianness on the Internet.

Within such a large Internet population as Asia, there are connections to an even broader Asian diaspora—to Asians who live outside Asia and retain a physical or symbolic relationship with their homeland. This sort of imagined community can also be characterized as "virtual" (Fernback, 1999, pp. 204–211), as it is experienced largely over the Internet. This has been examined in studies by Gajjala (1999), Mitra (1997), and Mallapragada (2000) of the Indian diaspora in cyberspace, but again these studies focus on how Indians living in the United States build virtual online communities.

Those who live neither in the United States nor in Asia are particularly marginalized within Internet studies by virtue of their relative numbers and invisibility. They are on the peripheries of both the Asian cultures from which they have been dispersed and the predominantly white, Western cultures to which they have migrated. Yet the Internet is often instrumental in maintaining the links between the different communities of which they are part.

Moving the focus away from Asian online communities in the United States, at least to those in other Western countries, is warranted if we are to explore patterns of use that might emerge across geographies of the Internet. Thus, in Western countries with large online populations, are Asians as prolific in their Internet use as in the United States? If so, why? Are they demographically or culturally similar or different? How might their minority status within a host society influence this?

This reorientation of perspective not only counters the U.S.-centricity of the study of virtual Asianness but offers an antidote to the once-dominant narrative about the monolithically white, Western profile of the Internet. It suggests a rethinking of center-periphery relationships in cyberspace as those who are often relegated to the non-white periphery appear to be infiltrating the not-so-white center. It is perhaps turning the lens away from one center of the West (the United States) and toward a multitude of centers.

Profiling the Online Asian

Despite their generally low profile, it is clear that the constituents of Asian online communities, irrespective of where they reside, are part of a "techno-elite" (Zurawski, 1996) who are privileged in their upward socioeconomic and geographic mobility, as well as their access to the Internet. Even now, few countries can claim that more than half of their populations are Internet users (ClickZ Stats, 2006b). Therefore, those who are part of the Asian online diaspora are still a minority within a minority.

If we look more closely at this group, a gender imbalance becomes evident, as male Internet users outnumber their female counterparts in every country apart from the United States and Canada (Pastore, 2001a); Greenspan, 2003a). Middle-class Asian men dominate the "techno-elite" of online Asia. This story is similar to the one heard in the late 1990s about the Internet's colonization by the white, Western, male middle classes (Apple, 1992; Johnson, 1996; Spender, 1996; Interrogate the Internet, 1996; Hoffman & Novak, 1998; Panos, 1998; Kendall, 1999). It seems that this is a narrative that has been repeated in Asia.

But let us inspect the nooks and crannies further, the margins within the margins, for those who are generally not visible in the metanarrative of online worlds or even within the plotlines of Asian cyberspaces: that is, the socioeconomically disadvantaged and women. When these factors are triangulated with Asian ethnicity (or any kind of ethnic minority status), the people who come into focus are characterized as "victims of the digital divide" (Nelson & Tu, 2001, p. 2), or the "roadkill" of the information superhighway (Nakamura, 2002, p. xii). In other words, their gender, class, and ethnicity immediately write them out of an online existence.

Yet all the evidence showing men's dominance of the Internet also suggests women are active online despite being in the minority. Likewise, women of color are part of virtual communities, as demonstrated in Gajjala's (1999) study of the South Asian Women's Network (SAWNET). Therefore, my argument is that when examining minority participation online, we must look for alternatives to the perspective of dominant majorities. This allows stories that would otherwise not be heard to be told. It facilitates insight from the periphery into the center, and from the local microlevel to the structural macrolevel. It provides an indication of how minority groups negotiate structures in which racism is embedded and which are not amenable to change. Technology is often instrumental in this, as those for whom a technology was never meant constantly reinvent it, appropriating it for their own needs, resulting in new cultural practices and politics.

My particular interest in the online activity of atypical users stems from my being one of those Asians who is missing from the above statistics, residing in neither the United States or Asia, but in the Antipodes, a woman of Chinese heritage who is part of the "techno-elite." When I lived in Britain, I worked with an even more unlikely group of Internet participants, who might be described generically in the words of Terranova (1996) as "the terminally unskilled and unemployed, single mothers on welfare, the old and the poor" (p. 72). Certainly, most of these descriptors applied to each of the fifteen ethnic minority women I taught as part of a foundation course in new technology. All resided in East London, an area that is subject to ongoing investigation because it is underdeveloped and deprived compared with the relative affluence of the Southeast of England (Rix, 1997, p. 119). East London's long history of immigration means that its ethnic minority communities have been particularly prone to high unemployment because of racial discrimination in the labor market and lack of English language skills in addition to economic disadvantage. This situation is exacerbated in key boroughs; for example, the 1991 census showed that more than 42 percent of Newham's population were non-white or of ethnic origin, of whom more than half were South Asian (i.e., Indian, Pakistani, or Bangladeshi). Black ethnicities include black Caribbean and black African. Unemployment is exceptionally high among Bangladeshi (more than 40 percent in 1991) and black African (nearly 40 percent in 1991) ethnic groups in Newham and Tower Hamlets (HMSO, 1991).

All the women were over the age of twenty-one (up to fifty years of age or older), and most were of "U.K. black," "African," or "Caribbean" ethnic origin as stated in the categories given on their application forms. However, Lippard (1992) argues that "a vital, sensible, and imaginative vocabulary can only be self-generated during the process of self-naming" (pp. 167–168), so when they were asked to articulate their ethnic identity, all had a different conceptualization: British-born Nigerian, Nigerian-born black African, British Asian born in Pakistan, English-born/Jamaican origin, British-born Jamaican, British-born Afro-Caribbean, African-born black, British citizen, Asian Bangladeshi, British Caribbean, British Afro-Caribbean, African/Caribbean, black African, black British, and Chilean Latin American.

Note that only two of the fifteen identified as "Asian." Rosie, who was born in Pakistan, and Noori, who is Bangladeshi, found themselves to be a minority within the wider peer group. Yet in the particular locality in which they lived and studied, both Rosie and Noori were part of a South Asian ethnic majority. This discrepancy between the demographic profiles of the local community and of the student group of which Rosie and Noori were part emerged from difficulties in recruiting Asian women to the course. This is certainly at odds with the notion of the technologically predisposed, superconnected Asian and suggests that there may be specific hurdles to Asian women's access to technology-related education and careers.

Rosie and Noori

Given the shared experience of class and gender between all the students, in what ways were Rosie and Noori distinguished from their peers through their Asian identity? Perhaps these factors can illuminate the particular challenges to attracting more Asian women onto the course and the reasons Rosie and Noori were able to overcome these challenges.

Rosie and Noori typified the disruption that Asian identity presents to the traditional black/white dualism of racial politics (Modood, 1997, p. 157). Apart from both self-identifying as Asian, both were Muslim and seemed to feel their minority status quite profoundly, not only among the student group but even within their own communities.

> When I go back to Pakistan to visit my relatives I feel like a stranger in the country that was once a part of me. To them back home I am British, because of the way we think perhaps we are too Westernized to them back home. To British people, I am still a Pakistani, British passport or not. (Rosie, personal interview, 1998)

> I came here when I was five and have never been back and really have no ties with Bangladesh...even though I may be Bangladeshi, I wasn't brought up like a typical Bangladeshi. I haven't been back to Bangladesh in twenty five years. Any proper Bangladeshi goes back every three to four years. (Noori, personal interview, 1998)

There is a sense of being neither here nor there, in or out: Rosie's migration from Pakistan to Britain has put her in an impossible position between "where you are from" and "where you are at" (Gilroy, 1990/1991, pp. 3–16). Noori says she is Bangladeshi but not a "proper" or "typical" Bangladeshi, and only part-time British. Even within her own ethnicity, Noori feels marginalized by her failure to return regularly to her symbolic homeland. This is not unusual in postcolonial identity politics, but it does seem to differentiate Rosie's and Noori's experiences from those of their peers, many of whom were British-born and thus one generation removed from the effects of migration.

At the same time, there is a reluctance to be typical, given the disadvantages they have faced as females in their own families and within the wider Asian community. Both have had their formal education cut short and apparently did not have a choice in this. Again, this was not apparent in any of the other students' experiences.

> I was taken out of school at the age of fifteen ... there have been obstacles in my way since I started this course; there are some people who have said I will give this up but I am fighting with myself and with the situation so that I want to be able to continue. (Rosie, personal interview, 1998)

> My childhood was not a happy one ... I suppose I got conditioned into thinking that all Bengali families were like mine ... Being the only girl I would get the least of everything: food, clothes, toys ... Bangladeshis seem either less intelligent than everyone else or do not realize that a good education will lead to a better future ... Even now any Bangladeshis who have computers, it is because a son is studying. If the daughter is studying then such an expensive bit of equipment is not considered important for her education. The Bangladeshis consider it wasteful to spend very much on a daughter's education because they will marry her off and she will settle down and have a family, never using her brain for studying ever again. (Noori, personal interview, 1998)

Rosie and Noori were acutely aware of the opportunities they were denied as young women and especially determined to seize the opportunities available to address this previous lack. This steely resolve emerged from the long wait they have both had to endure to return to study:

> Personally I have been out of education and out of a job for such a long time, I have devoted all this time for my children and especially having them in big gaps made me—It's like throwing away twenty years—I would say throwing away because although you are there for your children you are not contributing—like you can't bring any money into the house and then you have more problems. (Rosie, personal interview, 1998)

> Basically the last 13 years I have had to put a hold, a stop to my education because I have been bringing up my children and now my youngest is in nursery I feel I can go ahead and carry on with my education. (Noori, personal interview, 1998)

All the women on the course were mothers with child care responsibilities; Noori was among the youngest, but she had the most children. She was also in the minority in terms of being one of the few who was not a single parent. However, by her own admission, her husband was wholly unsupportive of her studies.

> I have got four children and a husband who doesn't lift a finger; he would sit here all day. I am in the middle of doing an assignment and he will see I'm on the computer but he has to ask for a cup of tea. (Noori, personal interview, 1998)

Rosie, the single parent of three children, stresses the "normality" of her situation. It was perhaps in this regard that she felt "one of the group," as the vast majority of students did not have a partner who was cohabiting with them and sharing the responsibility of child care.

> Before I began the course I was just a normal housewife with three children and the normal problems that every person experiences, especially when you are a single parent, and sometimes I think I have had more than my share of the problems, more than I can cope with and I think I cope pretty well. I try to cope with life so I am just an Asian, Pakistani Asian, lady, with one grown up son, a teenage daughter, and one other boy. (Rosie, personal interview, 1998)

In contrast to Rosie's insistence on her ordinariness, Noori emphasizes what has set her apart from others:

> I grew up with totally English people. There were no other Asians, no Afro-Caribbeans, no Africans, nobody with the same skin color as me. I was the only black girl there. In secondary school I was the only black person in that school. I grew up with two identities; at home I was Bangladeshi, at school I was British. (Noori, personal interview, 1998)

Noori and Rosie share Asian identity, although their experiences of this identity are quite different. Rosie is attached to her Asianness to such an extent as to feel that she has "lost a part of [her] life" in being labeled as British by relatives and friends in Pakistan. Noori seems to value her Asianness less, willing to discard it and identify as "black" instead: she refers to herself as black to illustrate her isolation in growing up with "totally English people" and with "nobody with the same skin color." That is, Noori expects to find affinity with others who have had similar experiences but who may not necessarily be Asian.

As Brah (1992) has pointed out, the term "black" has been contentious in Britain because of its inclusion of people of Asian descent, when many British Asians do not consider themselves to be black (pp. 127–128). Noori is unafraid of such controversy in affiliating with the black community in Britain rather than the Bangladeshi, Muslim, or Asian communities.

Anchorage and Liberation

Rosie and Noori exemplify the heterogeneity inherent in Asian ethnicities. In Rosie, there is a lack of certainty that leads her to seek anchorage points in her ethnicity, to seek solace in being "just an Asian," Pakistani, and Muslim. In Noori, there is a desire to escape the suffocating constraints of Asian Bangladeshi female identity. But both construct their ethnicities broadly so that they are represented through their connection with others within a diasporic community. However, they negotiate this differently, with Rosie grounding her ethnicity through her place of birth. That is, where she was born, or perhaps as Gilroy (1990/1991) puts it, "where you are from," is consistent with how she represents herself. Such geographical anchorage is a comfort and offers relief from the burden of identity politics.

Noori positions her ethnicity in another way, securing it to "where you are at" rather than specifically in terms of country of birth, nationality, or geographical region. Perhaps this articulates Giddens's (1991) argument that identity is now divorced from a fixed sense of place because of the role of technology in mediating social relations (p. 2). However, Noori's distancing from her Bangladeshi origins does not seem to be a product of being part of a virtual community. Rather, there is a mistrust of technology and its capacity to serve her needs based on previous experiences of media representation:

> TV has never portrayed Bangladeshi women as positive images. The only time I have seen them on TV is when they are showing victims of flood and the women are crying because they have lost everything. The only time I have seen British Bangladeshi women on TV is when they are shopping!
>
> There's nothing on TV, there's nothing in the magazines. The only thing I saw was for jewelry for Asian women on the back of a bus or something. Earrings or something...
>
> ...there are [Bangladeshi] women working both at home and the factories on the industrial sewing machines. The majority of Bangladeshi women are married off but if they do work they seem to be in low-skilled jobs, shop assistants, receptionists, threadpickers. The association with these technologies led me to think that Bangladeshis seem either less

> intelligent than everyone else or do not realize that a good education will lead to a better future. (Noori, personal interview, 1998)

Noori speaks as one of the "victims" of the divide between those who control technology and those who are oppressed by it. Where the media is concerned, this chasm is evident in the extremes of what is portrayed: the luxury consumerist West versus the disastrous helplessness of the East. It is also manifest in the disparity between the image and reality of the daily struggles of Asian women to meet the demands of their families. Noori's criticisms of the affluent, consumerist images of Asian women are based on their discrepancy from her world: how can Asian women be depicted as having the time and money to buy luxury items such as jewelry when their existence centers on making ends meet?

Thus Noori's introduction to the online world must be read in the context of her deep-seated cynicism about what technology can offer her. She neither expects nor hopes the Internet to alter the limits of her existence for fear that she may be disappointed yet again.

> I would probably find lots of Asian Web sites set up by professionals or academics dealing with things like Asian films...to marriage bureaus. For some of them there might be women involved but there would be a high ratio of men. There is a disturbing increase of Muslim extremism and I expect they will probably have a site somewhere. (Noori, personal interview, 1998)

Noori anticipates that the same gender hierarchies she has experienced in the media will be reproduced on the Web. This seems astute given research showing 72 percent of Asian men using the Internet, compared with 60 percent of Asian women (ClickZ Stats, 2001). Furthermore, with Asians shown to be particularly active online, and general users of the Internet increasingly contributing online content (Greenspan, 2004), both online use and production can be deemed to be male biased where Asian representation is concerned (Nakamura, 2002, p. 115). Because of the preponderance of male users of the Internet, Noori sees that the representation of women on the Web will be constructed for the "male gaze"—for example, as future wives for Asian men. It will be another extension of the ways in which women are depicted in the broadcast media (Jones, 1993, p. 253; hooks, 1992, p. 62): present in the image but absent at the level of both production and consumption. In other words, Noori believed that the content of the Web would not be intended for her, that she would be overlooked.

Noori's pessimism about the possibilities of the online environment contrasts markedly with Rosie's association of technology with enterprise and productivity:

> In the early sixties, a lot of Asians were going into the grocery retail business. Or open a restaurant and take-away ... Also they have gone into ventures such as into clothing factories, textile or property businesses. The younger generation is very aware of the computers, they also have much more freedom in making choices in educational and personal matters. So I suppose it's a mixture of new technologies associated with my ethnicity.
>
> I think I am very good with technology come to think of it. In twenty one years I have been married, I have only had two fridges. One I had, new one when got married it lasted eight to nine years. The second one I have still got; I bought a new one and my hoover only a second hoover. I look after my things.
>
> I am fascinated by the computer. I had only heard of one computer, that was ERNIE [the computer that picked government premium bond winners in the United Kingdom] ... I think I could relate to computers and the people who operate/program it. Our brain is like a clean sheet of unprinted paper ... Just like a computer we have to store information in our brains which works like a CPU. (Rosie, personal interview, 1998)

Rosie considers technology to be enabling when it is managed appropriately by people. The value of the technology is proportional to the degree to which one seizes the opportunity. She sees computers as having the potential to change people's lives. Therefore, as humans, we can exercise control over our own destinies by thinking like computers.

Rosie's optimism in relation to technology compared with Noori's cynicism provides some insight into how they might approach cyberspace and what might happen when they are given the tools of the techno-elite.

Connecting You Now...

Each student on the foundation course was supplied with a desktop computer and an Internet connection to allow flexible modes of studying away from the university at times convenient to the student. For most, it was their first contact with a computer and with the Internet. Therefore, there were steep learning curves to be climbed in acquiring both computer and online literacies simultaneously. Such challenges did not prevent Rosie from finding a wealth of resources, particularly in relation to her ethnicity and identity:

> There is a lot of information (about twelve categories). And three hundred and fifty four sites for Pakistan. There was a lot of information for all generations. For younger generation there was a lot of information on Pakistani rock groups, their music and on their backgrounds. I was surprised to know that there is so much information.
>
> I found Indian film stars, the film industry. It gave information about the film stars. Especially the ones you are interested in you can find out about their profiles. They give you their height, what their favorite food is, everything. I think I am more into music and that's why I went to this site and then I tried to learn if I can learn Arabic because although I am Pakistani but we are connected with being Muslims...they actually have lessons and things that you can download...You can learn any language and then I went to a site because my younger son was bullied at school...I went on word search and put bully or bullying and I have so much information on—or even some of the famous stars they had been bullied and what you can do about it, they give you the numbers to contact and sometimes they give you a questionnaire to answer back on some sites but I haven't filled in any questions. It really is great...there is so much information you don't know what to choose. It's a big market. (Rosie, personal interview, 1998)

Rosie does not seem to encounter any obstacles to joining the techno-elite despite her "minority within a minority" status. Not only is she part of a ethnic, gender, and class minority online population, but at the time of the research, there were fewer than 6.4 million households in the United Kingdom with Internet access, which translated to 27 percent of the UK population, or 13.5 million people (Pastore, 2000). Home connections were a relatively new phenomenon in Europe compared with the United States, and they were mostly the preserve of those with incomes of £15,000 or more (Pastore, 2001b). In this sense, both Rosie and Noori defied the conventional means of participating in computing and the Web, as they were on the wrong side of the economic digital divide. Also, whereas members of ethnic communities have been shown to spend less time online because they generally do not own home computers (Morrissey, 2003), Rosie's and Noori's occupation was effectively to explore computers and the Internet:

> On the Internet, there is so much information you can get carried away. You can find out about music from your country, about your culture, about laws, you can even get food manuals and there is so many things. You can try to learn the language and sometimes bring on the sound thing, it will say the words for you. If I could afford it, I think I would love the Internet. You feel the world is so small with the Internet. (Rosie, personal interview, 1998)

Rosie felt a closer proximity to the part of her life she had "lost," and also to all that she has wished for but never had access to. The Internet brought it to her fingertips, but she acknowledges it as a privilege that she would not have if she were not on the course: it is a technology that brings one close to others, but it is also out of easy reach. Rosie displays a sense of being lost among the sites she visited because of the array of material available relating to Pakistani identity. Her disorientation is illustrated in the difficulty of deciphering the trail of sites she visited. While cyberspace envelops her into a virtual network, it just as easily confuses her sense of place.

Similarly, Noori's foray into the online world produced lots of material but was disappointing in that it did not produce any resources that were specific to her identity position. That is, despite the amount of information available, when other dimensions of subjectivity are taken into consideration, there is not much on offer at all. The notion of community is fallacious: "I couldn't find anything because I am Bangladeshi and I am a woman and there's nothing" (Noori, personal interview, 1998).

Noori seems to consider herself unrepresented because of the intersection of her gender and ethnicity. Also, as a Bangladeshi woman, her gender and ethnicity become invisible or subsumed under more general Asian or national referents. Thus, it was impossible to nominate a particular Web text as most representative of her ethnicity: "There was not one site that summarized my ethnicity...I couldn't pick just one and say that represents me!" (Noori, personal interview, 1998)

Despite the volume of texts available, Noori was alienated by them. Whereas Rosie found this to be comforting because "you don't even feel alone," Noori was annoyed by what she anticipated and discovered. She saw evidence of more of what Nakamura (2002) identified as "cybertypes," racial stereotypes in online environments, particularly eroticized representations of Asian women on the Web:

> I guessed that there would be a lot of marriage bureaus and there was. When we went into it there was a whole lot, I think about ten or twelve lists of mail-order brides and stuff like that...I was searching on HotBot and first I typed in "Asian Bangladeshi" and all I got was about the population and ethnic minority in Britain and the rest was just mail-order brides, marriage bureaus, something for single men and stuff like that.
>
> None of these Web sites were aimed at women, they had men in mind and weren't bothered if women accessed them or not...apart from the one written by women, the rest are not going out of their way to cover issues important to women. (Noori, personal interview, 1998).

Noori seemed more disturbed by what she didn't find than by what she did find. Among the plethora of sites visited, the diversity she sought was not available. Perhaps it is indicative of a crisis of representation: the heterogeneity of the Web does not necessarily mean that there is a niche for everyone. Where minorities are not active in online production, the possibilities for networked mobilization are diminished. Online consumption does not a community build.

Conclusion

When an unlikely couple of novice computer users who do not own their own computers are suddenly thrust into the world of the techno-elite, there are as many contradictions as possibilities. While there is potential to participate in Asian virtual communities, to feel close to and part of something familiar, there is also the capacity for claustrophobia and discomfort. In the relationship between technology and ethnicity, gender and class interrupt the smooth transition between the online and offline. Being women in a socioeconomically deprived region of London exacerbated Rosie and Noori's minority status, not only within the South Asian digital diaspora but in the wider cyberspace. Focusing on minority activity online helps to avoid getting trapped into dominant narratives about the "digital divide," in which Rosie and Noori would generally be depicted as "victims." Indeed, far from bring helpless, Rosie and Noori manage and leverage their participation in cyberspace to the advantage of their families, and without the need for home computer ownership. At the same time, the research also showed that the Internet does not offer an easy escape from the offline world, no matter how much this is wanted at the outset, and despite the pervasive notion that it immerses and liberates: Noori's desire to be released from the constraints of her Bangladeshi identity is met with the disappointment of finding Web production to be male dominated and repetitive of what can be found in other media.

Rosie's and Noori's forays into South Asian cyberspace are examples of what can be seen and the stories that can be told when we inspect the peripheries, and, as in a Magic Eye picture, look beyond the obvious. In a Netizen's world, this means focusing on a localized situation (Kendall, 1999, p. 57)—communities outside the United States or the West—and interrogating how these intersect both online and offline (Mitra & Cohen, 1999, p. 269; Wakeford, 2000, p. 38).

Notes

The phrase "victims of the digital divide" used in the title of this chapter is from Nelson and Tu (2001, 2); the term "techno-elites" is from Zurawski (1996).

1. The names of the interviewees have been changed.

CHAPTER TWO

The Role of U.S.-based Indian Diasporas in the Entrepreneurship and Globalization Processes in the IT Industry

Venkataramana Gajjala

THE HISTORY OF ECONOMICALLY driven modern Indian diasporas dates back to the nineteenth century, when mercantile capitalism gained access to Asia[1]. This opened up enormous trade opportunities and eventually led to the creation of massive trade surpluses that were subsequently invested in mines and plantations in the British, French, and Dutch colonies in Asia, Africa, Latin America, and the Caribbean. Consequently, indigent workers migrated in large numbers to these colonies to meet the increasing demand for labor. This migration was made possible by the assimilation of peripheral economies into the rising world capitalist system, the improvement in transportation and communication, and the opening of the Suez Canal (Jayaram, 1998). For instance, in the case of Britain—the leading colonial power—migration was spurred by a declining birthrate and the abolition of slavery during the mid-nineteenth century. Excess labor demand was generated by an ever-expanding colonial economy.

According to Tinker (1993), three distinct types of Indian emigration occurred during the colonial period. The first was the migration of indentured labor—primarily from north India—that began in the 1830s and involved the signing of a contract by the laborer to work on plantations in the colonies. The second type was the migration of the kangani (an anglicized form of the Tamil word *kankani*, denoting "foreman" or "overseer") to Ceylon and Malaya and of the maistry (the Tamil word *maistry* denotes "supervisor") to Burma. The job of the kangani and the maistry was to recruit labor from rural areas in the former Madras Presidency, without any binding contracts (i.e., it was free under the law). The two systems were essentially the same except that the latter "was characterized by a gradation of middlemen-employers (the labor contractor, the head Maistry, the charge Maistry and the gang Maistry) and the innumerable deductions" (Laxmi Narayan, 1998, p. 7).

The system of indentured labor was abolished in 1920, and the planned export of indigent labor was abolished in 1938. The third type of migration, which followed the ending of the first two, consisted of the migration of merchants from the west and the north of the country (Gujarat and Punjab) to South Africa, Kenya, Uganda, and Tanzania and of those from the south to countries in Southeast Asia. The migration to East Africa also consisted of labor for the building of railroads. This type of migration was purely voluntary and was fully paid for by the migrants themselves; hence it was called "passage" or "free" migration.

There have been three main types of postcolonial migration: (1) the migration of Anglo-Indians—the progeny of unions between Indians and the British—to the United Kingdom and Australia, which unfortunately has been the subject of scant research; (2) the migration of skilled professionals to the United Kingdom and North America in the late 1960s and the 1970s; and (3) the migration of both skilled and unskilled labor to the oil-rich countries of the Middle East, which was temporary in nature and was driven primarily by the need to make some quick money to support families back home.

The second type of migration—that of highly skilled professionals (doctors, engineers, scientists, teachers, and more recently software engineers) to the United States—has been extensively researched, and deservedly so, because this migrant group has prospered and become one of the richest ethnic minorities. The principal cause for the migration of Indians to the United States in the 1990s was that the demand for talented software professionals in Silicon Valley and elsewhere far outstripped the supply. This gap was filled by low-cost skilled manpower from India—owing to the large number of university graduates with a working knowledge of English. Initially, this migration was considered a "brain drain" for India and a sunk cost, given the country's huge investment in the education and training of these professionals. As far as the receiving country was concerned, the debate concerned the alleged dislocation of native workers by these alien professionals.

A related, but more recent, issue here is the rise of the financially successful U.S.-resident science and technology (S&T) diaspora who have become part of a transnational entrepreneurial network (TEN) investing in India. Their important role in India's assimilation into the global economy has been enabled by the fundamental transformation in the make-up of the "Indian" diaspora since the 1990s and the process of economic liberalization set in motion by the Indian government around the same time.

Research on the forces of globalization thus far has focused mainly on the transnational activities of multinational corporations (MNCs) and has consequently ignored the expansion of S&T diasporas/TENs (Saxenian, 2002). Such networks are crucial because they share access to assets such as information, trust, and contacts (Portes, 1996). This chapter examines the reasons underlying the establishment of an entrepreneurial connection by these diasporas/TENs in their country of origin. It suggests that these diasporas/TENs may be as important a driver as states and MNCs of economic growth in their country of origin, thus leading to a transformation of the "brain-drain" process into a process of "brain circulation." It also looks at how the success of the U.S.-resident Indian S&T diaspora—especially in Silicon Valley—has created "spillover effects" that have enhanced the global image of India.

Entrepreneurship and Globalization Processes

Meeting U.S. Demand for IT Talent

It is appropriate to look at how Indian software firms initially capitalized on India's comparative advantage—its large pool of low-cost skilled manpower—by seeking low-end jobs such as writing code on the basis of specifications from major clients, remote maintenance of client software, coding only bits and pieces of a whole programming job, and price competition. This involved infrequent visits to the United States by Indian programmers for short-term onsite assignments, or what is commonly referred to in the industry as "body-shopping"—the export of "brainpower" at a price that exceeded the rate in India but fell far short of the going rate in the United States. This approach was then successfully aped by other Indian software firms and body-shopping businesses floated by the Indian S&T diaspora by employing Indian programmers with H1-B visas sponsored by them (Kapur & Ramamurti, 2001).

A significant gain from this was the rapid enhancement of the skill set of Indian programmers from on-the-job training in the world's largest and most mature market for IT goods and IT-enabled services. Furthermore, the successful export of human capital to the United States afforded Indian software firms the opportunity to move up the value chain, and as the wages of Indian software professionals in the United States went up, much of the work began to be outsourced to India. As this process gained momentum, the problem of remoteness—how to link the U.S. client with the Indian vendor effectively—became acute. This gap was bridged by the Indian S&T diaspora by creating professional and social networks to connect with their peers in India and thus marshal information, knowledge, talent, and resources.

Some members of the Indian diaspora took advantage of information and communication technologies (ICTs)—which have enabled not only small and medium-sized firms but even start-ups, regardless of their location, to globalize—and returned to start their own businesses; others, having established business relationships in India, commuted regularly between India and the United

States. Thus Indian-born, U.S. educated and trained IT entrepreneurs in Silicon Valley, for instance, are promoting new IT companies or floating subsidiaries in India. By virtue of their experience and contacts from being part of these professional networks, they can quickly identify prospective markets and react to them soon enough to reap the advantages of being first-movers.

Firms in developing countries get round the problem of being isolated from the highly profitable markets and the technology centers of the West—both of which are prerequisites for innovation (Hobday, 1995)—by opting for joint ventures. But this by itself may not be adequate; TENs can help with the diffusion of innovations and provide a way for firms in developing countries to globalize their operations. Economic reforms undertaken in many developing countries as part of the wider process of liberalization, in addition to the uncomplicated access to increasingly intricate ICTs, have accelerated the process of globalization. It should be noted, however, that ICTs by themselves cannot achieve the desired objective, the transfer of technology; in the final analysis, it is only TENs that can enable smaller firms and even start-ups in developing countries swiftly to capitalize on potential market opportunities—especially in the hi-tech sector—and flourish.

In addition to using India's comparative advantage in leveraging low-cost software talent for U.S. businesses, TENs have opened up the domestic (Indian) market—which is potentially huge—for U.S. firms, leading to the globalization of the Indian economy while simultaneously improving the U.S. balance of trade. Thus, TENs consisting of Indian immigrants in Silicon Valley and elsewhere have been responsible for technology-driven growth in their country of origin.

Likewise, adoption of free-market policies by governments alone cannot ensure the global success of firms in developing countries. Also, efforts by governments to nurture entrepreneurship—for instance, duplicating the Silicon Valley model—are highly unlikely to succeed in an extremely competitive environment because they are not a surrogate for the learning by doing and the sharing of insights that distinguish TENs.

For instance, it was TENs—along with domestic entrepreneurs—who sensed the expanding software development needs of several MNCs and seized the opportunity to provide these companies extremely price-competitive technical talent from India. They thus carved out a market niche for themselves and in the process emerged as a guiding beacon for the government of India, which realized that IT could be a stepping-stone to rapid economic growth.

Returning technology entrepreneurs have encountered such problems as the need for a reliable infrastructure but have circumvented them, for instance, by investing in power plants of their own or by buying power from companies in the private sector. Though business process outsourcing (BPO) is currently a major source of revenue for the Indian IT industry, it has become crystal clear to the returning technology entrepreneurs that in the long run they have to invest in more value-added businesses. This is corroborated by the fact that they have moved up the value chain and have simultaneously retained close links with the technology and markets of Silicon Valley. At this juncture, it would be relevant to look at two real world examples of TENs which demonstrate how Indian-born, U.S.-educated and U.S.-trained IT entrepreneurs have marshaled information, knowledge, talent, and resources to found and manage successful IT companies in India.

Diffusion of Technology and skills via Professional and Social Networks

MosChip Semiconductor Technology Limited. Established in 1999 with its headquarters in Hyderabad, India, MosChip manufactures fabless semiconductors. The company's Hyderabad design center, which employs more than eighty people (sixty of whom are engineers), designs and manufactures application-specific integrated circuits to support computing and network connectivity applications. MosChip acquired NetMos Technology, Inc., in July 2001 as a fully owned subsidiary and renamed it MosChip Semiconductor Technology, USA. The objective was to provide product intelligence and customer and marketing support from its Silicon

Valley base. In July 2003, MosChip entered the Internet protocol security space by merging with Verasity Technologies, the American developer of a highly integrated communication chip for last-mile Internet equipment.

Mr. K. Ramachandra Reddy (Ram), the chairman and CEO of MosChip, graduated from the Indian Institute of Technology in Chennai, India, with a BTech degree in electronics engineering and went on to get an MSEE from the University of Wisconsin, Madison. He spent twenty-three years working in Silicon Valley in the design, manufacturing, and marketing of integrated circuits. In Silicon Valley, he successfully started several semiconductor design companies, including Lotus Designs Corp., Silicon Logic, and Startech Semiconductor, all of which he later sold to larger companies. Ram worked for American Micro Systems in Santa Clara, California, before floating his first company in 1982. He was responsible for designing the world's first DSP chip. Ram's list of customers included PC manufacturers such as Compaq and mobile communication companies such as Nokia. Ram's career is a paradigm of the brain-circulation process—a successful Silicon Valley entrepreneur founds and manages a flourishing IT company in India and simultaneously retains close links with Silicon Valley.

*e4e Inc.*e4e is a U.S.-based IT company providing software engineering services, infrastructure and remote management services, and outsourced financial and technology services to global customers. e4e operates on the premise that it enables firms in the financial services, IT, and health care sectors to focus on their core competencies and build sustainable competitive advantages by deploying, managing, and operating their noncore but critical business processes more efficiently via cost minimization.

K. B. Chandrasekhar ("Chandra") cofounded and is currently chairman of the board of e4e. He is also the cofounder, CEO, and chairman of Jamcracker, Inc., an on-demand software company. Recognizing the vast capabilities of the Internet, Chandra founded Exodus Communications in 1994, which went public in 1998 and is considered one of the most successful initial public offerings (IPOs) of that year. In 1999, Chandra was honored as the Ernst & Young Northern California Entrepreneur of the Year. Fouress, Inc. was

Chandra's first entrepreneurial venture. The company—a network software design and development firm whose client roster consisted of big names such as Sun Microsystems, Adaptec, Toshiba, and Lockheed—was floated in 1992, and within two years Fouress had annual revenues of $1 million. Chandra moved to the United States in 1990 as country manager for Rolta India, Ltd. His duties included business development, marketing, and software consulting services for software developers and end users. His achievements at Rolta included adding big names such as Ford, DEC, ScanOptics, and Borland to the company's customer base. Chandra began his career in 1983 as a customer support engineer for Wipro Technologies, an Indian IT company, and moved up through various sales, marketing, and support positions during the seven years that he spent working for the company.

Today, under Chandra's leadership, the e4e group has expanded to include other companies, among them iSeva, Aztec Software, Vinciti AQ, and iCelerate, which employ more than 3,500 people and are in the business of providing IT and BPO services. In addition, the e4e group provides venture capital to Indian start-ups that it considers likely to be helpful to other group companies in achieving their strategic objectives. Furthermore, e4e has established Indian subsidiaries—e4e India, e4e Financial Services, and e4e Healthcare Services—and joint ventures in India, thus tapping into the vast and relatively cheap Indian software talent pool. Chandra symbolizes the process of "brain circulation" and is an outstanding example of how TENs have successfully duplicated the Silicon Valley model of entrepreneurship.

Unlike in the past, when India figured only as a source of low-cost labor and low-cost fixes, hi-tech firms in the United States now appear to be looking at Indian firms as partners in the design and development of new software products. The prospect of collaborating with Indian firms has become attractive because India is an emerging market due to its huge customer base. One excellent example is the rapidly growing wireless market in India, which could enable local firms to play a part in shaping the future direction of the technology and its application. An emerging market also generates profit opportunities for returning technology entrepreneurs; this is almost a prerequisite, since no entrepreneur is likely to return on the basis of promises made by the government alone. Furthermore, it is imperative that technology entrepreneurs return in adequate numbers (there seems to have been a positive trend in the recent past) and forge links with the local communities, since this is a precondition for the establishment of TENs. The government has to contribute by enhancing the quality of education and by upgrading the core infrastructure—power, roads, telecommunications, etc.—that facilitates and sustains the expansion of entrepreneurship-driven technology ventures.

In 1991, the government of India established the Software Technology Park (STP) scheme, and since then a big chunk of the growth in software exports has occurred in these hi-tech clusters. It is therefore appropriate at this stage to examine the microeconomics of growth and trade that has led to the formation of these regional clusters of entrepreneurship and innovation, which are comparable to Silicon Valley (Saxenian, Bresnahan & Gambardella, 2001).

The extant theories of clusters of innovative activity are based on two main premises. First, there is a direct external effect that is linked to the ability of companies to pull together information about the market and recent advances in technology more competently (Portes, 1996). Second, there could also be an indirect external effect resulting from enhancements in efficiency in the hiring of exceptionally skilled labor, in deployment of venture capital, and in "commercially oriented activities in universities or national laboratories" (Saxenian, Bresnahan & Gambardella, 2001, p. 5). One consequence of these direct and indirect external effects is growth in productivity, which provides a powerful incentive for start-ups to establish themselves in a cluster where others have already established themselves.

These external effects have two important consequences for economic growth in the context of agglomeration economies. First, by providing a favorable climate for innovation, clusters can considerably speed up the commercialization of valuable technologies; second, technology entrepreneurs, venture capitalists, and factors of production can achieve a much higher rate of return on their initial investment by being a member of such clusters (Saxenian, Bresnahan & Gambardella, 2001).

TENs have been an important driver of the growth of technology clusters in India which is in accordance with the *entrepreneurship and globalization processes outlined above*. This has resulted in spillover effects that have enhanced India's global image and have led to the development of a globally competitive software service industry in India. A brief historical review of the Indian software industry shows that although India was a latecomer to computerization, it has been one of the leaders in software exports in the past few years. The use of ICTs has led to a reduction in the optimal size of the firm and enabled direct sales on the world market at a competitive price.

According to the NASSCOM Strategic Review (2007), "Worldwide technology related spends are forecast to reach USD 2.1 trillion by 2010, growing at a CAGR of more than 7 percent over 2006–2010. Growth in global sourcing is expected to outpace growth in total spends, with up to USD 110–120 billion of the total amount spent on software and services in 2010, likely to be sourced through the global delivery model" (NASSCOM Strategic Review, p. 7). Key financial points of the NASSCOM Strategic Review are summarized in Table 1 below.

Table 1. NASSCOM Strategic review

Category	$ billion
Estimated IT services exports for FY 2007*	18.1
Estimated BPO segment exports for FY 2007*	8.3
Potential Export revenues from IT & BPO industries (excluding exports of software products) by 2010	60

*April'06–March'07

Further,

1. The IT-BPO sector is expected to earn revenues exceeding $47.8 billion in FY 2007

2. The IT-BPO sector is expected to employ 1.6 million people in FY 2007
3. The IT industry is expected to contribute 7 percent of GDP by 2010 (NASSCOM Strategic Review, 2007)

Conclusion

I conclude by examining some of the factors that have contributed to this spectacular growth, which has not gone unnoticed by the giants (IBM, Cisco, Microsoft, Hewlett-Packard, Motorola, Oracle, Sun Microsystems, etc.). These giants have established software centers in India to optimize their Indian supply chain or to perform R&D by taking advantage of India's low-cost high-skilled labor force. Simultaneously, leading Indian IT firms have gone multinational, too, by opening offices in the United States and/or acquiring U.S. firms to improve customer service to their U.S. clients on high-end projects or to keep abreast of developments in Silicon Valley—a process that was facilitated by the relaxation of rules regarding foreign acquisitions and the raising of capital abroad by the Indian government (Kapur & Ramamurti, 2001).

As mentioned, the connection between the forces of demand in the United States and the forces of supply in India was strengthened by the Indian S&T diaspora via the creation of professional and social networks to connect with their peers in India. Such contacts contributed often, for instance, to the appointment of a U.S. citizen of Indian origin as the CEO of a software development center established by a U.S. technology firm in India, or to the appointment of the CEO (again usually a naturalized U.S. citizen of Indian origin) of the U.S. subsidiary of an Indian software firm.

The success story of the U.S.-resident Indian S&T diaspora—especially in Silicon Valley—has created spillover effects of two types. It has widened the export market for Indian software firms to include many countries of the European Union, Japan, and Singapore, which in turn has led to an increase in the global demand for human capital from India and an expansion of the overseas Indian diaspora and its associated professional and social networks. It has also led to a demand for IT-enabled services from India such as accounting services (e.g., credit card administration), publishing services, audit compilation, tax return completion, medical record transcription, airline revenue accounting, insurance claims handling, business payroll processing, call centers, financial and human resource management, and engineering and design services. "Over FY2001–2006, India's share in global sourcing is estimated to have grown from 62 percent to 65 percent for IT and 39 percent to 45 percent for BPO" (NASSCOM Strategic Review, 2007, p. 3).

The U.S.-resident Indian diaspora, who are well represented in these areas, have used their contacts to outsource work to Indian firms. The Indian government contributed by initiating public policy that has included the establishment of technical colleges with English as the medium of instruction and by investing in communications infrastructure—especially high-speed Internet links and international gateways with sufficient bandwidth (UNDP, 2001)—thus leading to the informatization (Singhal and Rogers, 2001) of certain sectors of Indian society and the emergence of India as a world leader in computer software services (UNDP, 2005).

India is now a very attractive outsourcing destination for many of the world's leading MNCs, whose objective is to create global value chains. As stated earlier, the Indian outsourcing business owed its initial success to its ability to provide low-cost code-writing capability and program fixes to some of the world's leading MNCs, with the very convenient difference in time zones (*The Economist,* January 9, 2003).

Note

1. Earlier diasporas include the emigration of Buddhist missionaries during ancient times to central and eastern Asia and that of skilled and unskilled labor to the medieval Hindu kingdoms of Southeast Asia.

CHAPTER THREE

South Asian Technospaces and "Indian" Digital Diasporas?

Radhika Gajjala

PROCESSES OF GLOBALIZATION rely on a complex layering of discourses related to information technology, lifestyles based on the celebration of globalizing consumer cultures, and the seemingly contradictory invoking of national culture (defined through postcolonial bourgeois nation-building ideologies). The latter is exemplified in the discussion lists and Web sites that recreate and reconfigure religious and other cultural and nationalist utopian discourses (Gajjala, 2004). Examination of these digital diasporic discourses raises interesting questions regarding gender, class, caste, language, and race and their complex intersections. For instance, when is the "feminization" of particular kinds of labor perceived as a social and economic handicap, and within what larger everyday sociocultural and political-economic hierarchies do processes of "masculinization" and "feminization" become equated with power, domination, low prestige, or oppression? And how does the process of racialization in Western nations interact with the caste, class, and immigrant status of diasporic Indians within such hierarchies? How are layered and nuanced identities assembled in the tensions between mobility and immobility[1]?

In digital diasporas, encounters between the local and the global, the national and the international, and a multiplicity of representations occur. Such encounters, and the material and discursive shapings of community through these digital encounters, reveal, as Duncan and Gregory (1999) put it, "the politics of representation and spaces of transculturation...the continuities between a colonial past and a supposedly post-colonial present, and...the ecological, economic and cultural implications of globalization projects of modernity" (p. 1). Thus, Indian digital diasporas occur within racially, geographically, culturally, and socioeconomically marked configurations of the local, which in turn exists within power structures that regard a certain specific sociocultural, urbanized way of living as "global." Transnational, traveling subjects (whether they travel via "cyberspace," "road space," "air space," or even "sea space") negotiate online existence within such technological environments differently.

This chapter starts with an examination of the notion of virtual community and the false binary of virtual and real (In Real Life vs. Virtual Life). These concepts are pivotal to any discussion of online community formations through which Indian digital diasporas are manifested. I focus mainly on what appears to be the Indian diaspora in the United States (discussed in more detail by others in this book),[2] since a majority of Indian digital diasporic formations exist under hierarchies that appear discursively U.S.-centric even when the actual participants connect from all over the world. This U.S.-centric appearance of Indian digital diasporas comes from the implicit and explicit economic underpinnings of such digital diasporic formations. The prevalence of English as the main language and the consumer-culture signifiers rooted in multinational corporate branding also privilege U.S.- and U.K.-centric communication (even when the commodities have been manufactured in China, India, North Korea, or Indonesia). Thus a cultural U.S.-centricity is produced at the intersection of buyers and sellers of products, and the cultural significance of these products within the meaning-making hierarchies is produced through the interaction and information flow around global markets and brand names. This in turn produces value—socio-cultural and economic, for particular acts of consumption and social practices around this sort of consumption. The digital encounters of interactive meaning making in these digital diasporic spaces produce not only social and digital spaces of cultural representation but also contact zones of cultural contestation. In her influential book *Imperial Eyes: Travel Writing and Transculturation,* Mary Louise Pratt (1992) defines contact zones as "social spaces where disparate cultures meet, clash, and grapple with each other, often in highly asymmetrical relations of domination and subordination" (p. 4). The notion of a contact zone was originally predicated on the unequal power relations of colonial encounters. Such asymmetrical relations as those between the West and the East, the United States and India, and nonresident Indians and resident Indians are, however, still lived out in the age of postcolonization/neocolonization and the globalization of capitalism. In this chapter I briefly explore some aspects of the formation of virtual imagined communities of diasporic postcolonials online.

Virtual Community and Digital Diaspora: IRL vs. VL?

When home is no longer a concrete geographical place and exists within the "two-dimensionality of memory and nostalgia" (Karamcheti, 1992, p. 269), as is the case for the diasporic postcolonial, cyberspace may provide a way for disembodied minds (at both intellectual and affective levels) to make contact with apparently similar "beings," producing cyborg-diasporas[3]. Diasporas are linked to perceptions of some kind of "original" community that has been dispersed spatially and temporally, where the groups of people identifying with this "original" community re-member common histories, socio-cultural practices and locate themselves in relation to a real place.

In his article "Why We Argue about Virtual Community," Nessim Watson (1997) finds fault with the use of the phrase "virtual community." He argues that using the label "virtual" for groups of people interacting with each other online makes it seem as if the community "is not actually community." Watson writes:

> The term "virtual" means something akin to "unreal" and so the entailments of calling online communities "virtual" include spreading and reinforcing a belief that what happens online is like a community but isn't really a community. My experience has been that people in the offline world tend to see online communities as virtual, but that participants in the online communities see them as quite real. (p. 129)

He thus admits that online communities are different from those marked as offline, but he argues that the former are communities nonetheless. Watson also explores the gains and losses associated with using the metaphor of community for groups of people interacting online, arguing alongside Neil Postman that using the term "community" within online contexts may be inappropriate because online groups "do not contain the stake that exists in 'real' communities...they lack the essential feature of a common obligation. More accurately, online communities lack the consequences of not meeting or participating in the common obligation of most communities" (Watson, 1997, 122).

I would argue that holding "virtual" and "real" communities apart is unsustainable because online networks always have stakes in RL (real-life) communities. Communities are made up of group practices, discourses, and structures; virtual communities likewise are made up of group practices and discourses fully embedded in offline community formations, and the hierarchies in virtual communities (online networks) are embedded within these very power structures and ideologies. I use the term "community" in relation to online groups of diasporics because they do have a common connection, even obligation, to the "real" community that they are part of, whether or not they admit to these linkages.

Moreover, as Don Slater (Miller and Slater, 2000) suggests, the real questions perhaps lie elsewhere: "What is really required, therefore, is a move from asking about 'the nature' of online relationships and identities" to asking the entirely different question: "What do people do online?" (p. 539). Slater also calls for "more rich and integrated accounts of the social relations" occurring in online venues arising from "deep ethnographic studies of particular social groups with real histories" (p. 539). While differences do exist between an online interaction and the offline, between the spatial conditions in which one interacts online and the sharing of physical and temporal space talking to someone face to face or in a group, the question of what people *do* online shifts our examination to online community habitus (Bourdieu, 1984), which in turn is shaped by sets of online/offline practices from various geographic, cultural, social, religious, and political locations.

Thus, while it is true that the mainstream ideology behind the whole "global information highway" encourages the formation of niche virtual communities, where the participants try to distance themselves from the problems of what is perceived by them to be their lagging and backward country, it is also a fact that social relations and interpersonal exchanges within virtual communities cannot escape their connection with real-life political, economic, social, and cultural material practices. In Langdon Winner's (1997) view:

> The symbolic analysts of today's global webs of enterprise are shedding traditional loyalties, leaving everyone else to suffer in decaying cities....Such attitudes are found in 1990s cyberlibertarianism as represented, for example, in "Cyberspace and the American Dream" and in much of the hyperventilated prose of Wired magazine. These authors fiercely desire market freedom and unfettered self-expression with no sense of owing anything to geographically situated others. (p. 1)

The celebration of the freedom and the lack of boundaries of cyberspace assumes that cyberspace is a possible utopia for the privileged classes, separate from the reality of everyday suffering and deprivation that the less privileged of the world have to endure. Increasingly, cyberspace is marketed as a wonderland where gender, race, and all such markers of Otherness will be erased and melted down as we transform ourselves into texts and images online. This view, like the melting-pot ideology, fails to point out that the Others will be the ones who need to transform themselves into a Western-dominated information structure that is dictated by programming languages and Netiquettes laid out from a Eurocentric and U.S.-centric social, cultural, and political perspective. The lure of cyberspace, not unlike the lure of immigrating to the United States, promises a dream of freedom that becomes a possibility for the cultural/ideologically assimilated and materially privileged Other. The global village is performed when the Other has melted down and become a deracinated, degendered glob, and he or she is allowed the freedom to express his or her cultural difference within the "appropriate" postmodern, apparently multicultural paradigm. However, digital diasporas point to a more nuanced and layered existence in cyberspace than that suggested by either the utopic or dystopic views.

Whereas what is implicit in the utopic celebration of online community is the assumption that community formation starts with the individual and is rooted in the individualist rhetoric that pervades the technological imaginary, the dystopic view of online community ignores the pervasive everyday nature of our engagement with online sociocultural environments. Both these extremes treat Internet technologies as "new" and somewhat outside the everyday. The utopian and the dystopic visions of cyberspace and online technologies also overlook the fact that the individual is embedded within the practices, structures of power, and discourses that make up the community (Gajjala, 1998) and that these work through interpersonal negotiations and economic and cultural practices whereby inclusions and exclusions within such communities are determined.

So what is the difference between online and offline communities? I suggest that the difference arises in the engagement with the specific interface. By specific interface, I mean the context of the multiple technologies and at the intersection of the offline and the online. Roseanne Stone (1992) notes that "the actively multiple, situational and fragmented character of the online persona calls into question commonsense notions of the relationships between communities and the individuals that constitute them" (p. 6). In any community, the relationship between the community and the individuals who constitute it is established through continual negotiation and renegotiation of goals and ideologies—common or divergent. The relationship of the "I" to the "we," which in turn is situated in relation to a "they" and a context/ideology/culture/politics that the "they" positions the community with or against particular events. The community and individuals within the community, therefore, are bound by imagined and real accountability links. Structures of domination and hegemony within communities rely on reconfigured and reshaped networks of distributed power and redistributed accountability linkages—which both reinstate and reconfigure the status quo.

For the members of the community who are continually silenced or unable to express their experience or viewpoint within the given discursive framework or power structure, community is not a positive experience. If a member of a community is dissatisfied with the community, why not leave and form another community, you ask? This is definitely possible on certain social networks online, but when these social networks are heavily linked to offline economic realities and job-based teams (such as the many Linux communities that cluster together online in groups to work on open-source products), it is not altogether possible to "leave" and not always strategic to be silent.

Cyberethnographies

Information communication technologies, nationalisms, and religious diasporas are inextricably linked within processes of globalization. The processes of production and the cultural activities surrounding these processes are products of an economic globalization and transnationalization that rest on the need for self-contained identity formations (consumer demographics) and a performance of multicultural difference. Jihad and other religious fundamentalisms and nationalisms are examples of articulations of belonging and ways of imagining community that are being mobilized in the service of a particular vision of globalization.

As is the case with the processes of rebordering and the recent surge of ethnonationalisms in Eastern Europe and elsewhere, different fundamentalisms based in ethnic and religious identity formations are linked to emerging "global reconfigurations" that help the imagining of ethnic and religious communities transnationally while providing selective class-based access to global capital. Thus new hierarchies emerge that feed into "the logic of uneven global development." Sadowski-Smith (1999) further states, "It is essential to realize that...concepts of belonging are currently being mobilized in the service of larger political and economic demands associated with globalization"(p. 3).

What might be the role of virtual communities in fostering such nationalisms? Virtual communities are passage points for collections of common beliefs and practices that unite people who are physically separated. In the case of a diasporic individual for whom home is no longer a concrete geographical place, cyberspace presents itself as an ideal site for the recovery of community and connection with other diasporics with whom he or she explores and imagines a similarity of backgrounds. It has been suggested that cyberspace may provide a way for these disembodied minds to make contact with apparently similar beings. This production of digital diasporic identity at the interface of Internet technologies as online and offline intersect is determined in various ways by access to computer technologies, the design of these technologies, and the medium through which the identity is shaped. The collective imagination of the people involved will also be restricted by what they perceive as their material, social, cultural, ethnic, religious, geographical location.

An examination of the literature[4] dealing with the sociocultural, political, and discursive aspects of Indian cyberspatial formations reveals an interweaving focus on examining such online formations through theoretical concepts such as imagined community (i.e., how is the imagining of community online taking place?) and diasporic counterpublics. Within such framing, some attention is also paid to the structural and technical aspects influencing the sociocultural shaping of online spaces. It is important for us to examine these discussions for implicit assumptions.

There has been much discussion of the imagining of community in the literature on virtual community formations (see, e.g., Steve Jones's edited volume on *CyberSociety*, 1995). Imagining, as these explanations imply, happens on an individual level, where there is an attempt to connect the individual (often personal) experience with macrosociological features, often by translating one directly into the other. This is related to the imagining of any kind of community online on the basis of common interests, hobbies, collaboration on projects, professional interests, and so on. For instance, when we post to an online listserv community we imagine our readers/audience on the basis of how the listserv FAQs and information sheets describe comembers of the community. We imagine a kind of affective/intellectual communion. This imagining does not necessarily connect directly to our various real-life communities or to other imagined ones online.

The other sense in which the term "imagine" is used in relation to community corresponds to Benedict Anderson's (1991) definition of "imagined communities." In the case of the creation of virtual communities, this type of imagining applies framing around national, ethnic, religious, diasporic identity/subject formations. To quote Ananda Mitra (1997):

> The imagination that binds the members of the electronic group is the common memories of the same putative place of origin from which most of the posters c[o]me. The sense of community is based on an original home where everyone belonged, as well as a sense of a new space where the question of belonging is always problematized. Since the original home is now inaccessible, the Internet space is co-opted to find the same companionship that was available in that original place of residence. (p. 1)

Mitra suggests that there exist opportunities for various peoples in diaspora to form communities via the Internet, across place-based geographic boundaries, that are based on the constructs of commonality and fellowship while connecting to the conditions of existence of diasporic individuals. This is important in examining how the imagining of Indian digital diasporas has shifted over time.

Amit Rai (1995) and Vinay Lal (1999) extend discussions of online imagined communities to an examination of religious diasporas, specifically the Hindu diaspora and the discourses surrounding events in Ayodhya, India, in 1992. Still others have focused on the gendered nature of these online religious diasporas, with their implicit and explicit objectification of the Hindu woman as an icon of pure Hindu culture. Rai attempts to interrogate the diasporic publics and counterpublics in the context of Hindu religious fundamentalist activities. He too uses Benedict Anderson's (1991) concept of imagined community while arguing that cyberspatial nets provide a space for South Asian Hindus to construct and contest identities that are doubly marked by the nightmare of all the dead generations that we diasporics remember as India and by the always deferred promises of this new land of opportunity that is imagined as America.

Rai's (1995) use of the notion of imagined community leads him to examine the style in which diasporic communities are imagined and the regulatory fictions produced by officers of the British Empire. It is thus through the totalizing classificatory grid(s) produced in British colonial times that South Asian identities in the form of communal and religious diasporas are performed online. The performance of diasporic identities in these online communities is thus regulated through historical, political-religious discourses associated with colonial and postcolonial geographic territories and nationalisms.

Mitra (1997) and other researchers use the concept of imagined community implicitly in an effort to examine possibilities for the emergence of diasporic formations in varying degrees and seem not to question whether the Internet has the potential to enable a variety of liberatory and counterhegemonic coalitions. Lal, however, writes explicitly against the celebration of the notion of imagined communites online. He also begins to address the linkages between economic globalization, e-commerce, and these sociocultural diasporic cyberspaces by pointing to how the agenda of the Internet elites is linked with currently manifested hierarchies of globalization. Such a global economic climate, thus, suggests that rather than a panacea for the world's problems, cyberspace represents a more ominous phase of Western colonialism, the homogenization of knowledge and, in tandem, the elimination of selected local knowledge systems that fail to break into cyberspace.

Indian Digital Diasporas: Mobile (Gadget) Generations

So what about the generation of women and men who are growing up taking computers and the Internet as a given in their lives? Some refer to these as the gaming generation; I would like to call them the mobile (gadget) generation, since they move through the world in their own mobile digital aura.

There are several transnational venues in the digital diaspora that are inhabited by iPod-carrying, Game Boy-playing young men and women with their casual dress code and urban manners. Some of these spaces are less U.S.-centric than the previous Internet-based South Asian generations. Social network systems and blogs such as LiveJournal, Facebook, and hi5 (masked in semi-anonymity) blur notions of transnational sexuality as they hide behind Bollywood and other pop icons. There is a continuing play on gender and identity as the Bollywood icons produced in such communities are subjected to a gaze that blurs the boundary between heteronormative idolization of Bollywood stars and queer pleasure, while also producing uncertainty about geographic location as they appear to multitask between work, fun, and offline/online formations of friends. For this generation, being online is no more unusual than being on the phone. In fact, even the telephone is digital connectivity, as individuals incessantly text-message each other and download and exchange ringtones, pix, and flix. This elite group of young South Asians in digital diaspora are multiply literate and socioculturally flexible and mobile as they "hang out" in online communities of open-source developers, Bollywood and Tollywood fan groups, and so on. This Indian digital diaspora is producing new versions of elitism. The everyday practices of mobile generations in digital diasporas involve and invoke new and different kinds of problem-solving spaces from those inhabited by the materially underprivileged of the world. Thus, while the categories of "virtual" and "real" may not apply, we can certainly see that the socioeconomic and cultural gaps between the more mobile and less mobile are widening.

Conclusion

In this chapter, I have laid out some thoughts and observations concerning Indian digital diasporas in relation to the concept of imagined communities and to what has come to be known as the South Asian diaspora. I have argued that because these online communities are embedded in actual offline networks of diaspora and because the discourse and online presences/absences impact offline realities, the categories of "virtual" and "real" framed as a mutually exclusive binary cannot be applied. Furthermore, I have argued that the euphoric rhetoric about online community elides issues of digital divides and gaps between the more mobile and less mobile. As generations of Indians both in India and in diaspora continue to inhabit digital spaces through the use of mobile technologies in their everyday lives, Indian digital diasporas will be shaped and reshaped continually. This reshaping will always occur in an overall socioeconomic and political climate that is determined through access- and literacy-layered processes of globalization.

But these modern technologies can also help disrupt the fossilization of "Indianness" within Indian diasporic communities. It can challenge a fetishized notion of "an 'Indian' essence that survives all vicissitudes of changing material and cultural circumstances" (Karamcheti, 1992, p. 269). And this disruption will be forceful if there are counternarratives of present real-life material, social, and cultural conditions within the geographical region mapped out as "India," including, even especially, by Indians actually living in India.

Notes

1. In previous work (Gajjala, 2004), I have examined South Asian/Indian digital identity formations visible online. I have pointed to the interconnectedness of social, cultural, and economic factors such as IT work, nonresident Indian community formations and cultural practices, Indian national politics in relation to class, caste, and gender, as well as religious fundamentalisms and the ways in which political movements in India are influenced through nonresident Indian re-connections and re-investments and so on. Indian digital diasporas have come into existence within a time frame that includes the dot.com mania and post-dot.com distribution of transnational labor (e.g., call centers).
2. For instance, V. Gajjala's chapter on "The Role of U.S.-based Indian Diasporas in the Entrepreneurship and Globalization Processes in the IT Industry" lays out some economic and social factors that contribute to Indian digital diasporas.
3. "Cyborg-diaspora" is a term used by Karamcheti (1992) in her review of the book *Reworlding: The Literature of the Indian Diaspora* edited by Emanuel Nelson (1992).
4. For an overview of this literature, see Gajjala 2004.

CHAPTER FOUR

Sampling South Asian Music

Nabeel Zuberi

IN THE LATE 1990S and early 2000s, many hip hop, rhythm & blues (R&B), and Jamaican dancehall tunes featured South Asian instrumentation, refrains and voices, primarily from Punjabi bhangra and Hindi film song. These transplanted sounds included the percussive clusters of tabla and dhol, wistful flute melodies, melodramatic string arrangements, and the high-pitched vocals of Hindi film singers. Through digital sampling, these bits and bytes of South Asian music have been diasporic, migrating through networked media spaces to find new homes in black music. Some samples have appeared in several new compositions. The Deleuzian critic Drew Hemment (2004) terms this feature of the contemporary soundscape de-composition, "in which the museme or sound object becomes not just a question of looping or repeating indivisible units, but of reworking them in a nomadism where the nature of the musical fragment changes along with the territory it traverses" (p. 89).

This chapter examines some of the South Asian inflected black music, including recordings by Missy Elliot, Panjabi MC and Jay-Z, Truth Hurts and Rakim, Erick Sermon Ú and Redman, Timbaland and Magoo, and Bollywood Freaks. First, I listen to the songs to think through how the "black" and "South Asian" converse and sound together. These tracks deal mainly with the subjects of love and sex—the staples of R & B music—and so provide a site to investigate gender and sexual politics that juxtapose the "black" and "brown." I outline some of the issues raised about these pieces of music among fans, critics and musicians. These arguments centre on propriety, authenticity and the power relations involved in musical appropriations. Here I draw on academic writing, journalism, and online exchanges for the discourse about these sounds. The songs arguably include examples of Indo-chic, neo-orientalism, negrophilia, exotica, ironic or counter-exotica, the "rip-off" and the respectful homage "to my peeps across the color line." The conversation between different artists embodied in recordings in black Atlantic popular music idioms and the discourse around them reveal a range of cultural identifications. However, we cannot rely solely on a model of representation to understand the meanings of such musical appropriations. The qualities of sounds and their affects also motivate the art and craft of sampling and its appeal as a vital element in the aesthetics of contemporary popular music. In analyzing these technospaces we might bear in mind Deleuze and Guattari's (1987) suggestion that "it may be that the sound molecules of pop music are at this very moment implanting here and there a people of a new type, singularly indifferent to the order of the radio, to computer safeguards, to the threat of the atomic bomb" (p. 346). If this claim sounds utopian, these pieces of music, at the very least, persuade us to question stable identity discourses that too easily yoke musical styles to particular social groups. The mobility and mutability of sonic information in an environment of digital reproducibility also complicates any simple equation between forms of music and the racial or ethnic body. As Bruno Latour (1999) argues, "by following circulations we can get more than by defining entities, essence or provinces" (p. 20).

Missy and Jay-Z in the Bhangra Bins

There is a startling moment in the middle of the otherworldly music video for Missy Elliot's 1999 hit "Get Ur Freak On" just after Missy's gold and white jumpsuit instantly morphs into army camouflage gear at the very same time that she yells "Quiet." The groove is interrupted for a second. Her fellow dancers stop moving. Looking at the camera she exhorts the viewer to "Shshshshsh, hush yo mouth. Silence when I spit it out." The dancers also in combat gear return the "shshshssh" with fingers to their pursed lips. Missy digs deep and then audibly hurls some phlegm screen right towards her fellow dancers also in combat jackets. The saliva breaks up into pixels as it flies through this cyberspace as Missy commands "in yo face. Open your mouth, give you a taste. Holla!" The dancers holler back. Missy is extremely accurate with the transfer of bodily fluid. The cloud of moisture heads for one male dancer who opens his mouth. As the spit pixels agglomerate again, he receives the full amount, winces and then smacks his mouth with an ambiguous expression of pleasure and pain.

It may sound a bit like an incident of prisoner abuse at Abu Ghraib with Missy as diva of discipline and punishment. However, I appreciate this gobbing scene as Missy's lesson in digital dialogics. Utterances are exchanged in digital space. The material of that communication morphs across this media space. It is spat out, swallowed, and then presumably regurgitated in a new form. For good reason, MCs are referred to as "spitting" across a range of urban music genres. This is not the one-way gobbing of punk fans at their favourite bands on stage, but a chain of fluid exchanges. Immediately after this liquid transfer, Missy declares to camera:

"Ain't no stoppin me. Copywritten, so, don't copy me. Ya'll do it sloppily. Y'all can't come close to me." This challenge is readily taken up by scores of producers. "Get Ur Freak On" proceeds to be bootlegged and mashed up many times, rematerialized in freaky variations on white label vinyl and countless MP3s. At one end, Kid 606 adds glitches and jarring frequencies in a kind of violent electronic dissection of Missy's voice and Timbaland's beats. At the other, the faux seriousness of Go Home Productions' "Love will Freak Us" joins Missy's vocal to the sombre backing of Joy Division's post-punk hit "Love Will Tear us Apart." Skinny white suicide Ian Curtis becomes a goth backing-singer from the grave, at the service of Missy's foregrounded voice.

That such forms are often termed "bastard pop" acknowledges the discomfort in accepting the illegitimate progeny of call and response. While pop has continually eaten itself, the designation "mash-up" as a genre only appeared widely in the early 2000s. Critic Simon Reynolds (2004) describes the mash-up's appeal as "mild irreverence," and suggests that "the punk attitude to the form is as a kind of retaliatory vomiting of force-fed pop" (p. 306). But these mashups also show affection in their re-organization of familiar hit music. They often juxtapose sounds from musical styles, genres, and artists regarded as incompatible. "Love Will Freak Us" brings together the usually cordoned-off sound worlds of hip hop and doomy post-punk rock. We are encouraged to listen to the musical affinities between compositions in their new synthesis. Mash-ups make the familiar strange. Against Reynolds' desire for the radically new, can we not think of mash-ups as examples of a sensible media ecology, a kind of recycling that might be one of the defining logics of contemporary popular music? Repetition and multiplicity are central to the capitalist commodity logic of distributing value across versions, dub mixes and other remixes. Reynolds fails to consider the gift economy of mash-ups circulated without any desire for economic remuneration through peer-to-peer (P2P) networks. Most of these mash-ups have dubious legal status as they are traded free and bartered as MP3s across the Web and on burnt CDs.

The many lives of "Get Ur Freak On" depend on either Missy's rap or the instrumental groove being detached from the original. The metre of the voice and the rhythm of the music come from bhangra, so Missy's voice has been Indianized. She is not known as a rapper or MC with elaborate abstractions and metaphorical skills. Rather, Missy's signatures include her commanding catch-phrases and exhortations that function as hooks (e.g., "Work It," "Lose Control"), sexually explicit womanisms, and the pliability of her penetrating voice as it rides twitchy electronic rhythms.

On "Get Ur Freak On" the characteristic plucking staccato of a bhangra rhythm (provided, I think, by a looped sample of the string instrument tumbi) moves bhangra dancers' arms in the typical upper-body bhangra dancing that includes extending the arms above the head with wrists twisting in a *light bulb changing* style. The bass line and resonating drum machine frequencies designed by Timbaland also martial the lower end booty movements of bass-heavy music. He adds a tabla to the tumbi refrain to provide a rippling variation on the theme. The producer has been known to interrupt less productive studio time by going to strip clubs for musical inspiration. At these venues he is likely to hear those styles of R&B, hip hop, and dancehall, such as Miami bass and crunk, that come under the broad rubric of "booty music" (King, 2001). South Asians in the dance become, to quote Destiny's Child, more "bootylicious." While hip hop's hungry aesthetic does not just capture a Punjabi riff, it is changed by its incorporation. In its methodical way, the track modifies what I might call musical "South Asianness" and "blackness" in both music and dance. Reviews of dance club nights and my own participant observation register that shift in body movements.

I have not come across a negative appraisal of Timbaland and Missy Elliot's appropriation of bhangra. "Get Ur Freak On" is usually cited as the first major track in this cycle of Indianized R&B and hip hop. The reception of rapper Jay-Z's version of "Beware the Boys" has also been positive among South Asian Americans. Jay-Z (Shawn Carter) had already rapped over an "oriental" instrumental produced by Timbaland for one of his early hits "Big Pimpin." Over an Egyptian sample of a string section and flute melody that sounds influenced by imported Hindi films, Jay-Z adopted the voice of the hardest hardcore hustler of women: "You know I thug 'em, fuck 'em, love 'em, leave 'em/Take them out of the hood, keep 'em lookin' good. But I don't fuckin' feed 'em. First time they fuss I'm breezing. Talking about what's the reasons. I'm a pimp in every sense of the word, bitch." Such braggadocio helped to established Jay-Z's persona as a major player, hustler-cum-CEO, big spendin' hip hop mogul, the Roc-A-Fella, as he named his own record company (Berry, 2003; Smith, 2003).

On a promotional tour in Switzerland Jay-Z heard British DJ Panjabi MC's 2002 British and European dance floor hit "Mundian Ke Bachke" (Beware the Boys) in a club. He decided to rhyme on this alien beat. Panjabi MC sent Jay-Z the track, and he recorded a vocal with the support of The Neptunes, one of the most successful production teams in hip hop and R&B. This transatlantic collaboration was officially released in 2003. In this new mix, the sharp voice of singer Labh Janjua is still prominent as it calls out to a female audience "Mundian Ke Bachke" in the typical gendered call-and-response banter of bhangra from Punjab. That flirtation over the farmer's field or the urban dance floor is matched by Jay-Z as he becomes an orientalist lover man. He is a "snake charmer" who tells the Indian women ("mamis") to "move your body like a snake, Mama" as he approaches them. There is the charge of male competition as he rhymes "I'll take one of your chicks. Straight from under your armpit. The black Brad Pitt. I'll mack 'til six in the AM. All day I'm P-I-M-P. I am simply Attached to the track like SMPTE." With a pause on this celluloid simile, this voice of black machismo overcomes Punjabi machismo.

Little of the journalism comments on the sexual dynamics in "Beware the Boys." This may be due to an acceptance of this kind of masculine masquerade as part of the territory of bhangra and hip hop. It might also be evidence of a repression of anxieties around African American men dating Indian American women on college campuses and in city neighbourhoods. Tina Chadha's (2003) story about "Young South Asians" love-hate relationship with hiphop's new Indian beats in *The Village Voice* does not refer to the sexual aspect of Jay-Z's contribution. Rather, the U.S. Indian fans in the club express hopes that Black America has accepted their sound and thus made Indianness more audible in the mainstream media.

Another rap stanza in which Jay-Z throws off some lines critical of U.S. militarism gave the track extra credibility with liberal journalists: "We rebellious, we back home. Screamin' leave Iraq alone. But all my soldiers in the field, I will wish you safe return. But only love kills war. When will they learn. It's international Hov. I been havin' the flow. Before Bin Laden got Manhattan to blow. Before Ronald Reagan got Manhattan to blow." I cannot find this section in any of the MP3s I have downloaded, which might suggest to conspiracy theorists that it was removed due to political sensitivities. When I played "Beware the Boys" to an audience at a popular music conference in Wellington, one audience member noted that, unlike Missy Elliot, Jay-Z did not seem to have the flow to ride the bhangra rhythm successfully. This might mean that he gives up the rap after two stanzas for Panjabi MC's beats and Labh Janjua's voice to dominate the tune and sustain dance floor momentum.

Like many musical collaborations or guest appearances, "Beware the Boys" boosts the profile of both parties. Panjabi MC received some extra royalties and crossover exposure, particularly in the U.S. where he was relatively unknown outside the bhangra scene. Shawn Carter, Hov, or the Big P-I-M-P proved his nous for expanding his presence in other non-hip hop markets. This track came at a significant juncture in Jay-Z's career just before the release of *The Black Album* (2003), his retirement album. Jay-Z grandiosely stated that Jay-Z would soon be dead, as Mr. Carter would turn to concentrate on his business activities as a music company executive. In fact, what happened subsequently demonstrates that Jay-Z's presence in popular music took on the logic of sampling. His voice was widely distributed as he appeared as a guest feature on other artists' hip hop and R&B tracks, and their music videos. Authorial dispersal is common across these genres. Jay-Z popped up on his lover Beyonce's worldwide hit "Crazy in Love." While Jay-Z became Reebok's "ho," Beyonce made lucrative deals to "hustle" the products of L'Oreal and McDonald's.

Most significantly, Jay-Z's voice was put to purpose in the copyright wars, after it was sampled for *The Grey Album*, a collection of songs produced by Danger Mouse (Brian Burton). DM mashes up Jay-Z's vocals from *the Black Album* with the music of the Beatles' *White Album*. Music critics acclaimed *The Grey Album*, but EMI, the owner of the copyright to the Beatles' recording, demanded its withdrawal from the market and threatened legal action against anyone distributing it. In response, the music activist organization Downhill Battle called for a day of digital disobedience with Web sites and blogs offering the album for download on Grey Tuesday, 24 February 2004. So Jay-Z's voice was an unwitting combatant in the war for "semiotic democracy" with the mash-up on the frontline (Howard-Spink, 2005).

I again turn to "Beware the Boys." The real star of the track may not be Jay-Z's voice, Labh Janjua's vocals or even Panjabi MC's bhangra beats, but the bass line and intermittent electronic pulse sampled from the theme to the 1980s U.S. television show

Knight Rider. In 2005, this composition by Glen Larsen and Stu Phillips shared a new award from the American performing rights organization BMI for the most downloaded ringtone with Lalo Schifrin for his *Mission Impossible* theme. BMI (2006) estimated sales of ringtones in the U.S. that year as $500 million to rise to $600 million in 2006. As mobile sounds in networked places, ringtones are further manifestations of the logic of sampling applying more widely in popular music cultures.

After "Get Ur Freak On" and "Beware the Boys," the bhangra refrain has run through a flurry of remade hit songs. This group includes the Desi Kulcha's remix of Britney Spears' collaboration with Madonna "Me and the Music." The Panjabi Hit Squad, an Anglo-Asian production team, remixed R&B singer Christina Milian and Jamaican dancehall DJ Beenie Man adding bhangra timbres to their hits. Despite such interactions between South Asian and Black Atlantic musical elements, these respective musical-identity categories often remain quite rigid.

British academic Sanjay Sharma (2003) describes how one of his students at the University of East London, having fully absorbed the diasporic truism of "roots and routes" via Stuart Hall, Kobena Mercer and other black cultural studies scholars, went on to state that "Get Ur Freak On" demonstrated the roots of traditional Indian culture alongside the routes of black music's "changing same." In other words, musical blackness continues becoming while Indianness is frozen in an archive of unchanging traditions and types. This orientalist misreading sits alongside a measure of exceptionalism in African diaspora studies of music. Paul Gilroy's (1993) Black Atlantic gives space to South Asians to absorb the antiphony of black styles to re-articulate their now hybridised identities, yet there is no mention of a reciprocal relationship. As suggested by Vijay Prashad (2000), black composers, such as Duke Ellington, Jimi Hendrix, and John Coltrane, have been hugely influenced by South Asian cultures and incorporated their elements and techniques into their work. Gilroy's anti-anti-essentialism risks looping back on itself and not fully seizing on the implications of his statement that "calls and responses no longer converge in the tidy patterns of secret ethnically coded dialogue" (p. 110). In fact, music historian Ronald Radano (2003) argues that Gilroy's reformulation of Leroi Jones' "changing same" retains an ahistorical kernel that is "committed to a politics of center, to a transcendent, purely musical force that 'gets beyond' the instabilities of discursive contests" (p. 40). Radano rightly points out that "we simply cannot isolate a stable musical phenomenon from the historical matrix, as one might extract precious metals from ore or separate wheat from chaff" (p. 41).

The formations of black and South Asian music have penetrated each other. Bhangra has been "blackenized," to quote veteran R&B singer Hank Ballard's nationalist call to arms in 1972. Black style in music, sports, and fashion is a major component of United Statesian globalism. The ebonics of African American and Jamaican vernacular resound in local languages around the world. In its transnational dissemination hip hop has proliferated differences (Mitchell, 2001). Advertising and music videos sell hyperconsumption, once represented by the term "bling-bling" as a slice of the American dream or democracy. Soft drinks, fast food, sports clothing and footwear sales all benefit from the pan-branding power of Hip Hop, Inc.

At the same time South Asian cultural formations have also globalized. Bhangra flourishes as a transnational and networked music scene with significant nodes in northern California, New York City, Birmingham, Southall, Jullander and Bombay/Mumbai. Musicians, independent record companies, event promoters, bars, clubs, and retail outlets in diasporic communities have shaped local music scenes. Bhangra melodies and rhythms also resonate in Caribbean music styles, by way of producers in Trinidad and Jamaica and the long presence of South Asian diasporas in the region.

Bhangra's growth is also related to the territorial expansion of Indian media industries. Film companies or "banners," Zee TV and Bollywood 4 U are among the many operators that now distribute Indian content beyond the nation's borders. One major aspect of this economic growth has been the "Punjabification" of Indian popular culture as it globalized in the post-liberalization of the 1990s. Punjabi cooking is the dominant food in "Indian restaurants" outside India. Commercial films, made in the Hindi language at Film City in Mumbai, nevertheless mythologize the Hindu middle-class and patriarchal family values through representations of Punjabis and rural or small-town Punjab. The almost obligatory wedding song-and-dance sequence in these films is dominated by bhangra refrains. The North Indian cultural hegemony of nationhood is maintained as media companies appeal to the "homing desires" of Punjabi Non-Resident Indians (or NRIs). This diasporic market is one conduit for the hopes of Indian film companies eager to cross over into newer markets in the richer nations. Bollywood is the noisiest and most vivid element in a system that produces the affects associated with Indo-chic: colour, taste, subaltern suffering, spirituality, and exotic sexuality.

Lata and Asha in the Hip Hop Crates

Hindi film songs have been more vigorously consumed and reproduced in South Asian cultures than the movie features from which they spring. Apart from everyday singing, film songs appear in their entirety or in fragments in the mediated spaces of televised singing contests, music videos and film promotionals, cassettes, CDs, DVDs, MP3s, and the constant screening of Hindi commercial films in movie theatres and on television stations. Such is the currency of film song that many feature film titles are recycled from lyrics in well-known film songs from the past.

Alongside bhangra, the film song has become a rich source for samples. I now turn to the film song's recent digital mediation as its human voices and segments of instrumental music are sampled in hip hop and R&B. In particular, I want to listen to the way the female voices of singing sisters Lata Mangeshkar and Asha Bhosle have been processed and integrated into this recent black music. Lata and Asha's twittering high-pitched vocals have been ubiquitous in Hindi film song since the early days of playback singers in the late 1940s. Their stardom has involved the separation of their voices from their bodies. Through a cinematic ventriloquism, their vocals emerged from lip-synching actors. Songs could be released on vinyl under the sign of the singer, the movie, the actor, music director, or film director. HMV India owned Lata and Asha's studio performances and almost every piece of recorded film music until the 1980s. The dissemination of the audiocassette and the accompanying growth in piracy and independent music companies challenged the monopoly practices of HMV India (now Saregama) (Manuel, 1993).

Asha and Lata's voices have subsequently taken on a range of different modalities in emergent audio spaces. In the 1990s, the techniques of sampling and remixing crystallized into a new genre, the Hindi Remix. Film playback vocals were sampled for dance tracks on cassettes and CDs. Hindi Remix brought Indian consumers love of the melodic voice and lyrics from film songs together with the electronic and regular beats of disco. Indian percussion was often added to the metronomic beat of this European sounding dance music. The Indian consumer who found it difficult to dance to the offbeat and more supple grooves of African diasporic syncopations could now move to a more regimented beat with familiar lyrics and beloved voices. Pirates produced cheap re-recordings or remixes. HMV/Saregama effectively remediated its back catalogue and found a market in a new generation of listeners using higher quality cassettes and the new CDs. In particular, this trend revived the songs Asha had recorded with her ex-husband, music director RD Burman, film songs produced between the early 1960s and the early 1980s that had incorporated elements of American rock-n-roll, Cuban mambo, and German disco (Zuberi, 2002).

The recent digital sampling of Bollywood divas has scattered their voices further afield. One of the more celebrated cases of appropriation is the 2002 hit "Addictive" by female R&B singer Truth Hurts, which features a guest appearance by the highly reputed hip hop MC Rakim. A sample of Lata from the film song "Thora resham lagta hai. It looks silky" from the 1981 film *Jyoti* joins Rakim and Truth Hurts for a three-way. The producer of "Addictive" DJ Quik reports that

> I woke up one morning, I turned on the TV and landed on this Hindi channel and turned it up real loud. There was a commercial on, and I just got up and went into the bathroom and started brushing my teeth. I'm brushing and before I knew it, I was grooving. The beat on the TV was just in my body. I went back in there and looked at the TV—there

was a girl on there bellydancing, just like real fly. So I pushed record on the VCR. (Chadha, 2003)

DJ Quik's account is testament to Bollywood's exotic fascination with Arab culture, the globalization of Hindi films through networks such as Zee TV, and the technological ease with which one can sample anything from the recorded soundscape. Lata's voice is the first to appear on "Addictive" and a large chunk of the Indian recording is so prominent throughout the track that it sounds like a karaoke hand-clapping version of the song with English lyrics alongside the Hindi words.

Anand Bakshi's sampled lyrics consist of "Kaliyon ka chaman tab banta hai. Thora resham lagta hai. thora sheesha lagta hai. Moti larthay hai. Thora sona lagta hai." An English translation is "A garden of buds is made. It looks (or seems) a little like silk. It looks (or seems) a little like a mirror. The pearls crash. It looks a little like gold." The Hindi stanza is looped several times along with Lata's melancholic aah-ah-ah-aah and ohhhh-oh-ho-hoh sounds and a flute melody and tabla motif from the original.

Lata's voice is juxtaposed with Truth Hurts' English lyrics about an idealized romantic and sexual heterosexual relationship from the perspective of a woman: "He breaks me down, he brings me up. He fills me up, he breaks me down. He builds me up, he fills my cup. I like it rough. We cuss, we ball. We rise, we fall." The song describes a supportive partner who is also a ghetto-fabulous lover: 'My back is achin' From all love makin.' From head to toe, he makes me blow. He hits the spot." As Truth Hurts' vocal becomes louder and more urgent, Lata's moans could be taken as the exhalations of sexual ecstasy.

Rakim's voice enters at this point with a narrative about the dangers and threats of the gangster life when one sells drugs in the hood and goes to jail. He says he will quit this life for good once he's made "just another hundred mill." Truth Hurts' louder repetition of the chorus follows Rakim's declarations: "He's so contagious, returns my pages. He's got me anxious. He's what I've waited for. He keeps me guessin' Spontaneous. He's so persuasive. And I'm his lady." Now Lata's cooing in the background seems to suggest discomfort and loss. If male and female duets in R&B and hip hop explore the terrain of AfricanAmerican sexual politics, then the third Indian voice seems to offer more than one affective register, despite linguistically and sonically repeating itself throughout the song. Utopian heterosex and romance between African-Americans is materialized through the third voice of the Indian female.

Unsurprisingly listeners hear quite different things in the track. Their interpretive frameworks may be gendered, emphasize text or context, might appeal to the idea of self and other in radically different ways. These positions on musical authenticity can be moving parts in a conversation (Taylor, 2004). Here are two listeners, Tim and Geeta, in an online thread of discussion about the song. Tim writes:

> What I love about the use of Indian/Middle Eastern music on "Addictive" (see also: Foxy Brown's "Hood Scriptures") is how it steals not just from the specific sonics of the source, but also its fundamental performative alienness - the shivery drone of the sampled sigh and incomprehensible catewauling (sic) (at least from a Western perspective) are actually much more compelling in the service of POP than the by now familiar banghra (sic) riddims and flutes.

A few hours later, Geeta replies:

> You were right when you said that the Bollywood filmi song sample lended a fundamental performative alienness, but I take that further and say it's just plain haunting—more haunting for me, probably, than it is for you—you (perhaps rightly) are responding to the sounds themselves, the textures, how they work with rhythm. Growing up Indian and female in the US with Bollywood as an early influence, I see something completely different: I look at it from the calculated and yet vulnerable sadness in Lata's voice, love songs repeated forty thousand times over several decades in a thousand anonymous recording rooms: this sickeningly sweet soprano, a sample over twenty years old from a time when Mumbai movies were more "pure" and showed less skin. When it's displaced, looped, put into a modern song and a heavy beat is added, and layers are added of singers responding (and who can blame them?) to the sound and rhythm of the sample rather than its content or its original context, this rather fun, danceable song with the "cool foreign sample" ends up being, for me at least, very sad in a very strange way that I can't quite grasp. (ilx, 2002)

My own response to the track was that this African-American use of Lata's voice was funkier and more creative than that achieved by Hindi remix producers in India and the South Asian diaspora. Kevin Miller (2004) of the Department of Ethnomusicology at the University of California in Los Angeles notes another Desi listener's more ambivalent response: "When I listen to this I cringe the whole way through I love the idea, but I'm not necessarily thrilled with how they did it and how it sounds." "Addictive" is disturbing for sonic reasons that this listener cannot clearly articulate though s/he is not on principle against the use of Indian music in this foreign space. Further, Miller (2004) shifts his attention to the apparently more certain ideological ground of copyright violation. He argues that the failure of DJ Quik and the record company Aftermath to clear the license for this sample proves their ignorance about the size and importance of the Indian music and film industries. The lack of clearance could also be attributed to an ethical element of hip-hop sampling that maintains an open secret about source material so that producers that sample do not have to pay for copyright licenses (Schloss, 2004).

"Addictive" sold more than a quarter of a million copies in 2002 on producer Dr. Dre's label Aftermath, a subsidiary of Universal. Bappi Lahiri, the Indian music composer for Thora Resham Lagta Hai began a lawsuit against Aftermath and settled out of court for an undisclosed sum. Saregama Ltd. initiated a $500 million lawsuit that was also settled out of court. When I have mentioned this legal history to any follower of Indian film music they laugh since the so-called Disco King Bappi Lahiri is infamous for unashamedly and repeatedly ripping off whole swathes of sound from the Western pop pantheon without any copyright clearance. For example, many elements of Buggles' "Video Killed the Radio Star" have been lifted in the song "Koi yahan nache" recorded by Lahiri and playback singer Usha Uthup for the 1983 film *Disco Dancer*. Some of the critiques of sonic orientalism have been reluctant to discuss their occidentalist counterparts in India, delegating them to footnotes. By my estimation, Giorgio Moroder should be receiving handsome royalties from Lahiri and RD Burman's Indian disco tracks.

Despite the legal action taken against the producers and distributors of "Addictive" in the U.S., the Thora Resham Lagta Hai sample soon appeared in Jamaican dancehall in 2002's Bollywood riddim The Indian riddim, composed by Paul "Computer Paul" Henton of the In The Streets production team. The common practice in Jamaica is for at least a dozen artists to record different versions on the same riddim or rhythm. Lata's voice appears in all these tracks. These songs were released in a succession of singles and then compiled by British reggae label Greensleeves on the Bollywood album. In this context, Computer Paul added tabla and heightened the pitch of Lata's vocals and added reverb, making her sound more like a desperate Disney chipmunk's echo in outer space. Lata became a backing singer for an array of multiplying narratives about weed, war and slackness. Lata is one sonic element in the booty music of dancehall and its particular politics of gender and sexual display (Cooper, 2004). In JA, Bollywood joins other orientalist riddims such as Baghdad, Casablanca, and Diwali, most of which deploy Eastern motifs rather than directly sample from Arab and South Asian recordings.

The South Asian African-American interface is more troubling for the easy conflation of Middle Eastern and South Asian signifiers in some hip hop and R&B. Should we be surprised that this geographical and cultural ignorance seems to have been accentuated after the attacks on the World Trade Center and Pentagon? The music videos for "Addictive" and Eric Sermon & Redman's collaboration "React" both feature African-American dancers in West Asian and South Asian garb.

On the recording "React," Sermon brags about his ability to "scoop up an Arab chick" and attract women from several nations over a beat that incorporates what sound like syncopated harmonium stabs. The track repeats a sample of Asha Bhosle's vocal from "Chaandi Ka Badan," a song from the 1960 Hindi film *Taj Mahal*. This sample functions like a collection of short vocal stabs. Schloss (2004) has defined stabs as "quick, knifelike sounds [usually guitars or horns] that puncture the surface texture of the music before quickly receding" (p. 202); in this case, Asha's rhythmic burst of words: "Kissi ko khudkhushi ka Shauk ho To kya kare vo?

If someone has a fondness for suicide, what can one do?" Sermon blissfully responds to this backing vocal with "whatever she said, then I'm that." After 9/11, one could read this as a nod for suicide bombing, of which one might be surprised to learn that there are many ironic examples in recent hip hop. However, it is more likely that Sermon is clueless about Asha's words.

To Sample or Not to Sample?

Some black musicians reflect more on linguistic differences in their sonic fictions. Here women are still objects of desire but these articulations are not driven solely by the will to total male control. Timbaland and Magoo's 2004 single "Indian Flute" replays the scenario of two male MCs describing what they would like to do with an Indian woman. However, a South Asian woman responds directly to their approaches. "I love your Indian flute" they repeat umpteen times in the chorus. Her Indian flute? Could that be a reference to the clitoris?

There is a coy folky Indian flute lilt looped again and again for this fetishistic seduction. Tim and Magoo are initially somewhat hesitant: "Oh come here shorty, I like you/. What? You like me too. I don't understand a word you're sayin', but I'll talk to you." The object of their gaze is represented by the voice of Indian American singer Raje Shwari who sings mostly in Punjabi inflected Hindi, with a smattering of English. She is a "real" flesh-in-the-blood singer hired by Timbaland to sing in the studio, rather than a voice on wax processed through a sampler. When she teasingly suggests "zarasa choom loo to kya/Why don't you give me a little kiss," both MCs have to admit, "Well, I can't understand a word you're sayin." However, they do playfully attempt an ungrammatical line of affection in Punjabi: "Aaja soniye maine pyar kiya. Come on girl, I did love you." It is difficult to tell Timbaland and Magoo's voices apart as they take up rhyming alternately, speaking in tag team unison only in the chorus. Making rough sex not war, they both profess to "go slow & ill. Direct your every move like a young Benny Boom. After beatin up your womb. No need to put on perfume Baby girl, you leavin' soon." It is a fantasy record for black men and Indian women (possibly in threesome as suggested by the recording) to "get it on" despite South Asian and African-American racisms.

Though Raje Shwari sings live in the studio for this recording, Timbaland's production effects create the simulation of a voice heard on a mobile phone. The track evokes an idealized intimate conversation in the mobile's compressed tones. Alexander Weheliye (2002) has elegantly described the material sonic presence of pagers, phones and vocal effects in many R&B songs that explore sexual politics through these technologized modes of posthuman blackness.

If we consider the difference between the sampled female voice of Asha or Lata on vinyl at the mercy of the hip hop producer, and the embodied collaborations of musicians in the recording studio, one is tempted to argue for the latter "live" and often "face-to-face" creative dialogue as the way toward a more equitable politics of representation in popular music. Participation in hip hop scenes "on the ground" at the local level seems to offer more productive exchanges across ethnic boundaries than the "ripping" of a distant sound from vinyl. This is the implicit argument of Miller (2004), as well as Sharma (2005) and Maira (2002) in their work on hip hop and South Asian youth cultures in Oakland, California, and New York, respectively. The burgeoning but still marginal South Asian-American hip hop of DJ Rekha, Sammy Chand, Karmacy, The Himalayan Project, Feenom Circle, and many others intersects with the wider hip hop culture. With the word sacred, hip hop has been the pedagogical medium through which to reflect on power politics in particular cities, nation, and diasporas. The colonial histories of slavery, indentured labour, anti- and post- colonial movements, U.S. imperialism and immigration policy can open up new connections and ways of thinking and acting between subjects and communities locally and globally. For the children of immigrants born after 1965, hip hop has kept alive the examples of the civil rights movement and black nationalism (Melwani, 2002). Hip hop's diverse culture of music, fashion, visual media and performance also provides alternative life scripts for young educated South Asians who want to be creative workers and entrepreneurs in the cultural industries rather than the first generations, doctor-engineer-MBA triangle of model minority white-collar professional. Hip hop continues to generate defensive and offensive linguistic gestures and bodily codes with which to fight the emasculated Indian stereotype. Hip hop also teaches young people how to be consumers. As more than one wag has noted, alongside DJing, rapping, graffiti and breakdancing, shopping is the fifth element of hip hop. Even though South Asian Americans can also be seriously "into" hip hop music and styles, they keep their everyday encounters with African Americans at a distance due to racism and prejudice. They may not care about politics, but racism is a pretty common experience for Americans across the census categories.

Maira (2002) and Sharma (2005) concentrate on the important politics of identification with African-American culture. One of the reasons people love hiphop is its sound: the phat beats, the language and its flow, and the intense focused pleasure of hearing samples well integrated in a new cut. This doesn't mean that I wish to retreat into the conservative musicology of the music itself for my argument. Much of hiphop's appeal is not reducible to elective affinities between ethnic groups. Hiphop headz share a desire for certain types of sonic affect. We can examine the politics of representation in musical encounters between South Asian and African diasporas, but they don't tell us everything. One of the reasons for the sampling of Bollywood vocals and string passages is a broader hunger for high pitched sounds that stand out from loud drums and the low end bass frequencies of most hip hop and R&B music. As Douglas Wolk (2004) has explained, today's pop technoscape is dominated by the compression of sound—the evening out of levels in a piece of music so that the loud parts stay loud and the quieter parts become louder. He notes that these technological parameters can be used to advantage by creating

> dynamics based on off/on, sound-or-silence, especially if the sounds have very simple waveforms (synthesized tones, or heavily processed samples). That is the direction taken in a lot of the most popular hip-hop and R&B recordings right now: complicated arrangements whose component parts are very simple sounds and lots of space. (pp. 221–222)

The work of producers like Timbaland, The Neptunes and Kanye West demonstrates this tendency.

Sampling as a technological practice will remain central to hip hop aesthetics. The arguments around it testify to the many weaknesses of contemporary copyright regimes. On the one hand, sampled musicians should be renumerated for bits of their music that contribute to new compositions. On the other, we have witnessed the shrinkage of the public domain of cultural material that can be put to fair use and new creative ends. In any case, it is debatable how much credit should go to the original copyright owner of a fragment that has been significantly transformed. How do we measure the transformation of the sample in audio space? At the Future of Music conference in Washington DC in 2005, Hank Shocklee, producer with hip-hop group Public Enemy, posed this question when he asked a nonplussed George Clinton what one second of a snare drum from one of his recordings was worth. The music generated by sampling can also loop back to ultimately benefit the careers of the "original" musicians. For example, Hindi remix certainly raised Asha Bhosle's profile to the extent that she has a rejuvenated career working alongside the likes of Indipop musicians and U.S. Jart music groups like The Kronos Quartet. When one hears Asha's sampled voice sublimely sequenced with a broken sitar riff in DJ Vadim's production for One Self's hip hop track "Be Your Own," one has to acknowledge the creative imagination that makes something completely novel from old materials. This track is a digital transformation of the sounds that filter through Vadim's multicultural East London neighbourhood.

From "Get Ur Freak On" to Bollywood Freaks

Though the sampling I have discussed in this chapter has been lopsided in terms of African-American producers sampling South Asian voices and instruments, the countertendency is also rife. Beats and vocal interjections from R&B, reggae, and hip hop can be found increasingly in the music of South Asian producers in India and the diasporas. In their myriad ways these appropriations comment on the power of American black popular culture. While the politics of musical appropriations understandably vex ethnomusicologists, the relative autonomy of digital audio-spaces can generate its own critical rhetoric about South Asian-African diasporic relations.

A few years ago on one of my weekly visits to an Auckland record shop, I suddenly noticed a seven-inch single. The green cover was a grainy video capture of a female Bollywood star from the 1960s or 1970s. The shaped eyebrows and mascara gave the era away. The only other information on the front cover was the phrase Bollywood Freaks in red ink. The back cover of the single had Side A and Side B typed in the style of Hindi lettering. The A-side was "It's a Player's Holiday." Its flip was "Last Night Bollywood Saved My Life." There was no information about the artist(s), only an email address and a list of "big ups" to an incredible array of African-American, South Asian, British musicians and celebrities. The record inside yielded no information. It was a white label in the parlance of dance music's bootleg economy. The record was pressed on lurid green vinyl. The single as a material object reminded me of the coloured 45s of Hindi film songs that my parents played when I was a child in the UK. With these clues, I was sure that the musician behind this recording was a British Asian. "It's a Player's Holiday" turned out to be a mash-up that combined the instrumental of "Hooray! Hooray! It's a Holi-holiday" by 1970s Afro-Germanic superstars Boney M (still big in India) with Jeru Tha Damaja's vocal from his moralistic rap "Ya Playin' Ya Self." Jeru's voice declaiming women and men for their vanity had been sped up, making it closer to the pitch of Lata and Asha and undermining its pious intent.

"Last Night Bollywood Saved My Life" took what sounded like the Latin-tinged music of a cabaret number from a 1960s Hindi film and deposited the African-American female vocals of Indeep's 1982 dance anthem "Last Night a DJ Saved My Life" on to its surface. I found that Bollywood Freaks had recorded a subsequent single, another pair of mash-ups that mingled Bollywood with Black America. Bombay Gangstarr speeded up the mellifluous voice of hip hop MC Guru from the group Gang Starr to make him sound more like a playback singer in the vein of Mohammad Rafi. "Don't Stop till You Get to Bollywood" took Michael Jackson's 1979 disco hit "Don't Stop till You Get Enough" but replaced his relatively high-pitched voice with the Hindi vocal of Usha Uthup, a rare example of a deep voiced female in Hindi commercial cinema. The cover for the single was a red-tinted negative of a woman reproduced six times like an Andy Warhol silkscreen. In both these singles, the sound sources selected by Bollywood Freaks suggest affinities that already existed since the original Usha Uthup vocal was backed by an Indian take on disco. South Asian producers like Biddu created disco-pop hits for black, white and brown performers like Carl Douglas, Tina Charles and Pakistani duo Nazia and Zohaib Hassan during the 1970s.

The presence of so many "Indian-sounding" refrains in contemporary popular music genres in which the ethnicity of performers is not visible or difficult to read reminds us that ultimately music doesn't belong to anybody or any identity category. Rather, we belong to its flux of moving sounds, images and discourses. The technospaces of popular music are sites that articulate cultural differences through a range of wide gestures. Digital sound samples are used as ethnic resources unmoored from the bodies of performers. They are also used for sonic qualities that may not necessarily signify Indianness, South Asianness or any vague oriental tendency. Jon Hassell's description of "a sonic texture like a 'Mona Lisa' which, in close-up, reveals itself to be made up of tiny reproductions of the Taj Mahal" seems apt for the presence of South Asian samples in much of today's digitally produced popular music (Oswald, 2004, p. 135). However, the pixels, bits, and bytes constituting new sonic textures are likely to come from a wider range of sources than the singular Indian image of the Taj Mahal suggests. A million moments from the South Asian recorded archive will continue to scatter across popular music genres. Some of them will be recognized. Others will resound in alien contexts with little or no acknowledgment, but they will still be integral elements of emerging soundscapes and sonic subjectivities.

CHAPTER FIVE

On Purple Pleasures: Digitally Assembling Bollywood

Amit S. Rai

TO BEGIN in the middle of things, we could note that until very recently, cinema and its varied media technologies in colonial and postcolonial India were managed, organized, and categorized according to imperial or nationalist or capitalist pedagogies as a way to educate the nation's "unproductive expenditure" with its attendant popular unruliness and black-market economy. In this framework, cinema was a mass delirium that, paeans to Mother India or Raj Kapoor and Nargis in Russia notwithstanding,[1] was both in excess of national development and a possible technology of development. Cinema, then, is a kind of contagious practice that the Censor Board chief, excise commissioner, and state's chief medical officer could quarantine among them; simultaneously, cinema is one of the key sites and stagings of nationalist integration (Chakravarty, 1993). This colonial and postcolonial assemblage for the administration of Indian film culture includes the historical shift in the dominant design of cinema spaces from European art deco *moderne* (e.g., Eros, in Churchgate, Mumbai [Bombay]) to postindustrial multiplex (e.g., IMAX Wadala, Mumbai)—a shift in the frames of probability that the spaces of cinema offer to its consumers, which would also include the patterns of religious and social segregation of urban space. Moreover, we should consider the refunctioning of, first, theater, then radio, then terrestrial, cable, and, finally, satellite TV with the cinema's audiovisual commodities—transformations in different cultural apparatuses becoming mass entertainment. Of equal importance is the folding of cinema into various computer interfaces and across technological platforms, which has both necessitated and facilitated the globalization of film finance as well as changes in censorship legislation and practice from "censorship" to "certification"[2].In terms of consumption technologies, the rise and fall of fan clubs and cinema societies is part of overall shifts in the policing and deterritorializing of cinema/new media space through changing technologies of social hygiene, once centered on norms governing subjectivities in the social space of the talkie, to mechanisms of constant modulation of consumer habituations in VCD culture and Internet cafes. My itinerary in this chapter proceeds through contagion, with one thought affecting another, affecting another, until a kind of critical "mutagen" is created. I focus on the deterritorialization of sensation, audiences, and genre in contemporary Bollywood narrative in the context of a "potentializing" reading of two versions of the cinematic classic *Devdas*.

As should be clear, India's new media assemblage draws its commercial force from the image of film culture as contagion, as an exuberant devotion infecting bodies by their promiscuous mixing. Contagion, in fact, is one of the modes of assembling this open whole. We should note, then, that the idea of contagion implies something crucial for Hindi-Urdu film criticism. For instance, consider two important castings of the "assemblage" of Bollywood film: Madhava Prasad's (2001) exploration of Mumbai cinema's mode of "heterogeneous manufacture" (Marx, 1867) and Priya Jha's (2003) more recent recasting of the interruption of "gender difference" into the formulaic narrative of genre films. First, Prasad:

> If we consider the Hindi film is conceived ... as an assemblage of prefabricated parts, we get a more accurate sense of the place of various elements, like the story, the dance, the song, the comedy scene, the fight, etc. in the film text as a whole ... What makes this method of functioning unsuitable for Hollywood is the fact that a material substratum—the story—is the point of departure of the production process and its transformation into a narrative film is the final goal of that process. (p. 43)

Note that in Prasad's "mode of production" analysis of film, production is prosthetic, and cinema's assemblage is an always already preconstituted addition of already assembled parts, where one part (e.g., song) is sutured to another (e.g., dance) by a cohering thread (i.e., story). This assemblage is a "quantitative multiplicity," discontinuous and spatializing, like 1 + 1 + 1. I will return to these properties of multiplicity below; for the moment, what I would like to highlight is that as one of the dominant forms of institutional analysis of Hindi-Urdu popular cinema, such mode-of-production analyses as Prasad's have the overall effect of presenting a cinema of static reproduction: reproduction of ideologies, narratives, identities, infrastructures, even prefabricated heterogeneity. In other words, Prasad's notion of Hindi-Urdu cinema's mode of production presents a "preestablished realm of possibility," more of the same (Massumi, 2001, p. 116).

In partial contrast, Jha's (2003) work, meticulously attentive to gender and difference, focuses on the interruptive qualifications of cinema's performativity. Jha argues that Prasad "gestures toward the recognition of the filmic system itself as the creator of cultural meaning espoused through the contents of the filmic space" (p. 49). In his model, she suggests, the diegesis of the film is often read as less significant to the reading of the entire film "as the narrative structure merely follows the formulaic demands of the industry"(p. 49). But no popular Hindi film can be comprehended without resorting to the formulaic device used to create national cohesion. Because of a liberal "borrowing" from Hollywood, Bollywood films often reflect the growing need of India to convey its entry into metropolitan modernity after colonialism. The Bollywood apparatus, as a highly charged institutional tool, joins past and future at threshold moments. In this sense, the interruption of the formula with genre draws attention to the artificiality and construction of history while allowing "difference" to be part of the formulaic narrative. This is where I think songs do a particular kind of historical work that cannot be translated purely as "musical" across national borders (Jha, 2003, p. 49). What is most striking, indeed compelling, about Jha's recasting of Prasad is her insistence that the apparatus undergoes a kind of potentialization at threshold moments: difference interrupts the static reproduction of genre conventions. But in signal ways this zone of potential is reduced to the possible-actual circuit of gender hierarchies, performative subversions of national territorializations, and melodramatic codes. Thus,

> the recuperation of the melodramatic mode as "feminine," the alternative triangle of the diegesis, the song space, and the spectator are always shifting and adapting themselves to the dictates of the spectacular vernacular of the cinema. In this sense, the erasure of women from the song space rather than the exclusion of women from the national imaginary instead reinscribes them even in their physical absence. In both *Zanjeer* and *Sholay*, the feminization of the song space attests to this fact. This is evoked through the reminder that melodrama, as part of the private sphere, the home, that women are seen to occupy, accompanies the circular narrative and claustrophobic environment of the song space. (Jha, 2003, p. 51)

Jha here centers the analysis of cinema, then, on the vernacular aspects of film culture. But what happens when we situate film in terms of continuous change across dynamic thresholds?

Despite their differences in style and framework, what both Jha (2003) and Prasad (2001) demonstrate is a discursive closure of the zone of potential of Bollywood's media assemblage and a foreclosure of the virtuality. This foreclosure enables modes of interrelations—becoming in belonging. This line of thought stands far from the depoliticization of film criticism yet claims its radicalization through the potentialization of the multiply mediatized, overdetermined body. Furthermore, the issue here is not merely the reduction of cinema's effects to representational or cultural codings but its inability to think the radically new, that is, the specific changing connectivities of popular cinema from the 1950s social through the 1970s homosocial melodrama to the 1990s diasporic deterritorializations, and the constitutive destabilizations and blurrings of these genres, the becomings that belong to the specific interval of interactions of this cinema's media assemblage: from black-and-white 1950s montage to the new algorithms and postproduction color adjustments.

Most immediately, this brings us to the work of Madhav Prasad, Ravi Vasudevan, Geeta Kapur, Tejaswini Naranjana,

Ashish Nandy, Abhay Sardesai, Rahul Shrivastav, Shilpa Phadke,[3] and Lalitha Gopalan. It is Gopalan's (2002) text *Cinema of Interruptions* that forms the most immediate backdrop to my work. There are moments in her study of action genres in Indian cinema when the thought of what I am calling media assemblages (contagious and continuous multiplicities) proliferates beyond the dialectic of interruption and continuity, moments when certain complicity is marked and made strange by becoming something else. Gopalan's text, more than any other monograph on Indian cinema written in the past ten years, pushes the terms of debate beyond their contemporary framings. She does this through the cinephiliac's odd and always risky combination of marking pleasure and launching critique—an interpretive pleasure in cinema that is profoundly ethical and itself interruptive. For instance, the beloved song and dance sequence, despite their profilmic integration into narratives (e.g., in the work of Mani Ratnam), do "not totally escape their ability to circulate separately from his films to other economies of production and reception" (Gopalan, 2002, p.129). This flight, traversing bodily and aural-imagistic economies, limns the connectivities of what I will call India's media assemblage. This separation establishes connectivity through constantly unstable feedback-loop relations across aleatory variables, technological-human substrates. This detachability of the song-video is an aspect of the production and consumption of the Hindi film music industry that from its inception in the 1930s has necessarily involved film music in volatile connectivities, even though, as Arnold (1993) argues, song and dance sequences were more often integrated with narrative flow in the first phase of film music, roughly from 1931 to 1965. Given the changes in sound recording and consumption such as Dolby in India, what needs to be considered is the actual shift in affective flows of intensive consumption from narrative to song and dance movement.

Thus the differences across media—say, from the cinema hall, to DVD, to cell phone ringtones, to JPEG/MPEG, to the Internet, not to mention from dialogue to song—could also be addressed as the qualitative shifts in the modulation of noise-desire-information-affect in media synesthesia. Crucial here is that, like many digital technologies, the shift into digital modulation happens at a level below experience; one definition of MPEG (short for Moving Pictures Expert Group), for instance, suggests that the MPEG algorithm achieves a

> high compression rate by storing only the changes from one frame to another, instead of each entire frame. The video information is then encoded using a technique called DCT[4]. MPEG uses a type of *lossy compression,* since some data is removed. But the diminishment of data is generally imperceptible to the human eye. (Webopedia, 2004)

In other words, how do contemporary forms of digital control (encryption) alter the nature of the body's functional connection to quickly changing media technologies at a level *below* or, better, *before* perception of the media event? Very simply, all this implies a fundamental shift in the sensations of cinema. If cinematic practices, and not just the cinematic *image*, are indissociable from the Internet, Dolby sound, or satellite TV, and if the form of *bodily attention*—exteroception, proprioception,[5] interoception, in short, affect—produced through the cinematic sound-image and its attendant practices, institutions, and spatiotemporalities has been supplemented by the new media, we must understand this phase transition since a qualitatively different dynamic threshold is emerging as dominant between these technologies and the active bodies of media consumers; thus the thought of cinema itself must become a diagramming of the body's sensations in and through this new media assemblage.

But whose body? And which media technologies? Two initial "method" problems present themselves. First, are you suggesting that all bodies, regardless of gender, religion, class, caste, race, sexuality, et cetera, are equally implicated in this new dynamic threshold? Second, are you assuming that satellite, Internet, cable, et cetera have transformed the totality of film-media culture in postcolonial India, where access to and knowledge of such technologies are themselves technologies of social and economic exclusion and control?

Keeping the above in mind, how can we think the transformation in Bollywood cinema's regime of sensation in the era of its new media assemblage as a qualitatively different kind of solicitation of the body's essential creativity, its openness as a center of indetermination—in short, its virtuality? How can we think the assemblage's specific form of power as a violent machining, as the reproduction and containment, or reframing, of deeply entrenched and repeatedly produced inequalities, clichés, and habituations across heterogeneous populations, a human multiplicity structured in dominance? This chapter attempts to hold these lines of critique in productive tension of virtualization and containment as the conditions of emergence of Bollywood's global media assemblage.

Let us return to our initial point of departure and pursue some founding elements of this new diagram of media, bodies, and power: Bollywood's biogram.[6] I argue that Bollywood, indeed any globalizing media, is an assemblage of interpenetrating multiplicities that establish partial and temporary relations of motion, sticky synergies, or pulsions with and through populations, bodies, image-sound regimes, technologies, spaces, commodities, affects, and capital (and this is never an exhaustive list). How can we characterize the nature of Bollywood's media assemblage in the present moment? We could begin by noting how the speed of Bollywood seems to be increasing and its effects intensifying.

The interrelated reasons for these phenomena include the following. First, because the media lines of dissemination are proliferating, Bollywood, and its connected media cultures, can be found all over the world, from Kenya to Iran to Trinidad to Kansas. The proliferation is in fact an intensification and refunctioning of an earlier mechanism of Bollywood distribution—the Gulf-states-oriented videocassette piracy and the informal, ad hoc circulation of videocassettes from the 1980s on. Each site is a singularity; for example, Shahrukh Khan in Port of Spain, Trinidad, does not have the same affect-image as for the diasporic community in Toronto, Canada. However, each site is also multiply connected to the globalizing circuits of exchange, consumption, and viewing that define contemporary Mumbai cinema's regimes of bodily pleasure. Furthermore, Indian production houses were only recently granted "industry status" by the government, thereby opening up new avenues of legitimate, or non-"black," financial underwriting for a small but growing minority of top production houses. Even when we acknowledge the creativity of active audiences in the moment of the body's implication in these assemblages, we must understand the overwhelming numerical strength of Hollywood. Way back in 1992, the United States was exporting around $3.7 billion worth of media products annually, compared with the European Union's $300 million; this dominance only increased in the subsequent decade. Hollywood's presence in India is strong and getting stronger by the day. With a potential audience of 50 million, Hollywood is committed to competing for market share in India, hoping for at least 10 percent of the market in the coming years. The Hollywood majors, of course, welcomed India's decision in 1995 to liberalize the import of foreign films and to raise the quota from 100 to 200 films. The government also agreed to the private negotiation of film imports, which, until 1992, had to be channeled through the National Film Development Corporation. The dubbing of Hollywood films into Hindi and south Indian languages has generally been a great success and has focused Hollywood's interest in multiplex exhibition. As Thomas (1999) notes,

> While *Jurassic Park* cost US$30,000 to dub, it grossed US$6 million in India. Furthermore, marketing "synergy" for each new release with multinational corporations like Pepsi and Kelloggs has inaugurated an era of intense competition for audiences. Hollywood is also looking towards building multiplexes in India. Time-Warner has explored the idea in Bombay but has met resistance from domestic theatre owners. (p. 285)

Hollywood is looking for synergy in Mumbai, not in the vein of cultural imperator, but more along the lines of parastic catalyzer (Pashupati, Sun, & McDowell, 2003; Rajagopal, 1993, 2001; Sonwalkar, 2001). Thus, Hollywood dominated economic models for globalizing media assemblage are spurring the proliferation and profusion of Bollywood's own media lines: 20th Century Fox recently produced *Ek Haseena Thi* (2004) starring Saif Ali Khan and Urmila Matondkar and directed by the new prince of Bollywood, Ram Gopal Varma. Varma calls his strategy pursuing the best of the West:

> There's going to be a massive change. A lot of old filmmakers are going to go out of business. Anyone who looks at a film as a formula of one song, two comedy scenes and three action scenes, who doesn't look at the totality of the film, is lost now. Anyone who follows the old prudish traditions, of showing a bush's shaking leaves when they mean people are f—ing

> behind a tree, is gone. And anyone who doesn't follow the West is gone. For many people in the business, their pride won't let them. But following the West is not surrendering. Following the West, the best of the West, is following originality. Western innovation is superior, and I think we're just beginning to understand that. With my films, I'm targeting the urban multiplexes, the sophisticated media-savvy young crowd. Frankly, I couldn't give a f— for the villages. (Pillai, 2003)

Unfortunately, rural people across India have to give a fuck, since Gopal's new cinema is helping to define the very consumerist regime that has sought to relegate rural life to some kind of vestige of arrested development (Jayaprakash, 2000). Indeed, the cross-merchandizing pleasures on sale at the new Bollywood multiplexes both in India and in the diaspora are mapping ever more territories of bodies and population for ever more speedy dissemination, but through very specific community and subjective controls and modulations Bollywood accelerates also as the types of media-lines multiply: digital, satellite, air, print, Internet, radio, optical cables, DSL, and telephone wires. We should note further that both these accelerations in media speeds are discernible at the level of the diegesis and cinematography: faster narratives, fewer songs, shorter average shot duration (Bordwell, 2002), long- to medium-range shots enabling quicker and more economical editing, a shift from mass lighting to spot lighting, new pop-formatted music videos, and so on. Just as these changes are crucial, and contradictory, they are discernible in the profusion of genres, subgenres (*Dil Chahta Hai* with *Jogger's Park* with *Mango Soufflé*), and minimarkets (Chor Bazaar [Mumbai] or Hamidia Road [Bhopal], as black or gray markets for Bollywood memorbilia), and through the thinking and extension of surfaces of creative friction in acting, directing, cinematography, choreography, musical direction, lyric and dialogue writing, and the integration of different technologies associated with each of these elements of production. All these forces establishing relations of motion and difference between and within themselves generate new forces of pulsion, hence the change in the very nature of Bollywood in multiple becomings.

The thought of the becoming-media of cinema through the habituated modulations of the proprioceptive body and processes catalyzing contagious multiplicities would trace the diagonal flow of desire that traverses the assemblage, constantly moving from one outlet to another, adapting Benjamin's (1999) take on Nietzsche's eternal return. Thus the becoming-media of cinema "is an attempt to reconcile the mutually contradictory tendencies of desire: that of repetition and that of eternity" (p. 166). In one interview, a serious Shahrukh Khan fan in New York City, a female college graduate in her twenties, a transplant from a small immigrant Hindu community in Minnesota, remarked that the city provided fans with so many more "outlets" to connect to Bollywood. For this fan, the Internet was a mode of connecting to the oceans of information on Hindi film, to fanzines, and finally to the affecting body of Shahrukh, who perhaps was the original "King of Bollywood":

> Q: How have you found that experience of Bollywood on the Web?
> A: I think it's great. There's just, there's a lot out there, you know. There's, there's almost too much out there, to a point where you know do you want to read about the stars, do you want to read about their love life ... I feel like you can just read everything now. Um, you can have chat room discussions if you want, though I've never done that. But it's like a whole new level that they've taken it to. Bollywood's definitely ... it's, it's kinda become like a lifestyle rather than a trend. Like, people are doing everything [through it]. You know, I subscribe to *Stardust* [...] Actually I don't subscribe to it but, because I've been called such a FOB [Fresh off the Boat] it was my birthday present from a friend of mine. (Q: Ah, but you kept it.) But I kept it. (Q: And you read it.) Um, the whole subscription ... So I get it every month now ... I mean, you can see all the pictures and the, you know, the photo shoots and, and whatnot. But, um, I was recently an extra in the *Kal Ho Naa Ho* movie ... the Karan Johar's movie with Shahrukh Khan, I did three scenes with them ... mind you, I don't think a lot of people would because they pay you nothing for an all-day thing in which all you do is, like, sit around and wait. You know, I don't know if you've ever done that thing, but you seriously just wait, and I was ... just there because the whole, like, aura of being around Shahrukh Khan was, like, so amazing. Like, he is one of the most intense people I've ever been around. Seeing him is definitely an experience. Like, you can see him on film, but to see him act in real life is, is definitely something. (Q: He's a great actor.) He is. He's a great actor and has, like, his expression, and not only just his face but, like, his whole body expression. He's really intense. You know and even just watching him, he makes you just feel, like, feel what he's feeling. You know, like,

there's a scene where he's sad about it and he's just crying and you want to go up to him and tell him that it's gonna be okay 'cause he just looks so sad. (Interview with "Anjali," October 18, 2003, New York City)

These moments—the Internet, *Stardust*, movie set, Shahrukh, affection—are not sequential stages, or lines of a narrative, but are implicated in one another, infolding continuously and taken to a new level of intensity. The meeting, for instance, of the Internet-friendly digital cinematography (which would include within its own assemblage both the technology and logos of T-1 lines, Lucasfilm THX, and miniDV) and Bollywood melodrama has produced many skins, many relations of motion, many affects. All of them in one way or another signal the death of Bollywood, which is not to say the end of Bollywood: both Benjamin (2003) and Derrida (1982) have reminded us that what is dead wields a very specific force of the new. The futures of Bollywood are being born within the different fetishisms specific to this cinema—that is, its own fetishes, which include light-skinned virgins, Switzerland, muscular *jawans* (male youths, soldiers), colonial nature, the workerless city, the cosmopolitan consumer, and the global logo. In the West was seen the fetishism of middle-class European Americans dancing "authentic" bhangra in the aisles during the screening of *Lagaan* at the Film Forum in Manhattan. The proliferation of these fetish surfaces—the skins of an assemblage—guarantees a future, monstrous or not.

The acceleration of this cosmopolitan commodity, the globalized Hindi film, has also accelerated the production of these skins, surfaces, plug-ins, *hat ke* folds,[7] and amplitudes from Jackson Heights (Queens, New York) to Jahangirabad, Bhopal. Film literally touches just about every aspect of South Asian life.[8] It will immediately be asked: To what political end? What are the politics of a contagious multiplicity? Luciana Parisi (2004) and others suggest that what is at stake is the thought of a feminist sexual agency—a nonhuman, open, mobile, multirhythmic becoming. I agree with this characterization, but the forces of globalized homogeneity/standardization and the proliferation of transnational difference compel another thought as well: privatization and subnationalisms are both "contagious multiplicities" beyond their spatializing quantifications of human multiplicities; they also generate and function through affectivity, rhythmic durations, and qualitative differences. In that sense we can see that what is contagious about Bollywood is also what can be habituated in practice and for greater and greater profit. Here is a short yet annoying answer: Bollywood's media assemblage does not have an essential politics, but it is always and everywhere political.

Indeed, that productive expenditure is still the dominant mode through which cinema's cultures are managed in urban spaces in north India is no longer the question. Where are the movements and the differences taking effect? One difference is that before liberalization, cinema's music-image-sound regime (i.e., the image and musical style of cinema) was seen both as a possible enemy (hence ban rock n roll, but enable popular actor Shammi Kapoor to achieve stardom as Elvis's postcolonial double) and as a model citizen of the nation, its prodigal son, so to speak (Raj Kapoor and Nargis in Russia, for example). Today, into the open arms of a shining Mother India, dominant Hindi cinema has returned bearing the mask of the nonresident Indian. As I have suggested above, the proliferation of global consumerist sumptuary practices, tied to transnational travel and capital accumulation, in the past decade in India has been enabled by and has in turn spurred the convergence of film and new media and marketing in India. This new assemblage brings together film exhibition with transnational cross-marketing strategies. Posters with Hrithik Roshan, Ash-wariya Rai, or Aamir Khan, who are considered "A-grade" film stars, and Coke or Pepsi stand outside and inside the movie hall. The images and effects of new technologies double the bodily effects of digital sound, while Dolby logos and compact disks become objects of a modernizing desire. The connectivities of new media, such as the Internet, cable, and satellite amplify and thereby change the nature of star discourses and fan practices. In addition, the pedagogical temporalities center on facilitating different forms of consumption: shorter films; detachable, highly stylized song and dance sequences; the gradual elimination of the intermission, already beginning in multiplexes; the partitioning of screening times through an analysis of audience segmentation; the proliferation of liabilities and the consequent pedagogies of chances and risks of the present, nationalist-diasporic-primordial belongings.

In the past decade-and-a-half, the neat genre distinctions that divided Mumbai cinema into art films and commercial films have broken down. In the heyday of art films, roughly from the late 1960s to the early 1990s, social topics ranging from colonialism to feudalism, gender inequality, Westernization, and the tyranny of elite traditions formed the narrative whole in a cinema whose visual style was relentlessly realist and explicitly noncommercial. Movies such as Shyam Benegal's *Ankur* (1974) or *Bhumika* (1976), Basu Chatterjee's *Swami* (1977), Girish Karnad's *Godhuli* (1977), Ketan Mehta's *Holi* (1984) and *Mirch Masala* (1985), Mahesh Bhatt's *Arth* (1982), or Vijaya Mehta's *Pestonji* (1988) were self-consciously part of the "parallel" genre of art cinema, one founded in a clearly legible political and social commitment and a state-sponsored aesthetic—the long-take, realist montage and on-location shooting. In the postparallel cinema of the late 1990s, best represented by such works as Shyam Benegal's *Zubeida* (2001) and *Hari Bhari* (2000), Ram Gopal Varma's *Satya* (1998) and *Jungle* (2000), Mahesh Manjrekar'a *Vastaav* (1999) and *Astitva* (2000), Kalpana Lazmi's *Rudaali* (1993) and *Daman* (2001), Mani Ratnam's *Dil Se* (1998) and *Bombay* (1995), Vidhu Vinod Chopra's *Parinda* (1989) and *Mission Kashmir* (2001), Bollywood music sequences are folded into social critical narratives, most often in an extension, commentary, or deepening of the narrative itself. Combined with a narrative-governed montage logic, realist camera eye, and big-name Bollywood stars, this cinema signals the end of clearcut genre distinctions in Indian cinema. Bollywood cinema today has broken with its own laws of genre.

One can lament this break as an erosion of the "artistic" in the face of the massification of all South Asian culture through globalization. Another route through other questions is possible, however. If genre distinctions are breaking down between the relentless social realism of art cinema and the pure fantasy of the commercial film (and this genre binary was never as rigorous a distinction as it might appear in hindsight), what combinations of form, modes of address, and affects has this shift entailed? In one sense, commercial cinema—in its narratives, its visual style, even its choreography—has incorporated aspects of the former parallel cinema. What does this mean for the commodity form of Bollywood cinema? What kinds of visuality and aurality have functionally assembled in commercial cinema as a result, making it more or less commercial? What relationships circulate among fantasy-travel, narrative-interruption, and exhibition space in the new postparallel cinema? What is the work of music, specifically the detachable song-dance in this new cinema?

The dominant framing of these questions in mainstream Western film criticism has come to affirm the essential nature of Bollywood as a nonnarrative, sensation-based entertainment. This repeats and displaces a colonial trope of the animalized pagans reveling in the riotous night of unreason, from James Mill to Thomas Babington Macaulay to Winston Churchill and beyond. I suggest that Bollywood cinema is both nonnarrative and traversed by an abstract machine for the commercial production of affect. I argue that the human-technological assemblage of contemporary media is transforming the sensations of cinema toward new media horizons and new consumer habituations that fragment, interrupt, and jam the sensory-motor schema that enables narrative. This intensifies the human-media interface as a site for indetermination, or potentialization, and simultaneously deploys this indetermination as value-added immersion in such technologies as Dolby sound and the multiplex.

Take, for example, a recent near-hit in Hindi-Urdu cinema, one that translated Quentin Tarantino's *Resevoir Dogs* into a Bollywood idiom. *Kaante* is the all-male multistarrer that brought together on the same screen Amitabh Bachchan, Mahesh Manjreker, Sanjay Dutt, and Sunil Shetty. It was shot entirely in Hollywood, with "largely American" technicians (Kehr, 2002). Yes, certainly, "large" and "American" are good adjectives for this production, but what makes it Bollywood? According to Kehr, Bollywood embodies a "tradition of excessive generosity," in which "too much of just about everything" finds its way into the film (p. 40). Although he admits that there are a couple of "snazzily edited music videos," he writes that "scenes like this make you realize that Bollywood is essentially a non-narrative art form, in which coherent storytelling plays a distinctly secondary role to the purple pleasures of the moment" (p. 40)[9]

Narrative, or coherent storytelling, in this view is seen as marginal, secondary to Bollywood; it celebrates the "colorful" pleasures, which, through their fleeting nature, perhaps provide this form of criticism its favorite adjective for Bollywood—"ephemeral." Bollywood is the realm of the ephemeral pleasures, alternately known as "cheap pleasures." Even more "sympathetic" Western critics are invested in the discourse of the essential Otherness of Bollywood as a kind of excessive pleasure. For instance, Chute (2002) writes,

> Bollywood cinema's peak achievements, like those of Hong Kong, devote immense amounts of creative energy, ingenuity, and highly evolved craftsmanship to the life-affirming task of delivering intense pleasure to the largest possible audience. But while Hong Kong cinema works by radical compression, Bollywood operates by expanding in all directions. (pp. 35–36)

Of course, there is also an important strand of cultural criticism from South Asia and its diasporas that argues about the nonnarrative nature of Bollywood (Gopalan, 2002; Jalal, 2002, Mishra, 2001).

In what follows, I argue that Bollywood cinema draws criticism toward an interrogation of the very possibility embedded in narrative itself, that Bollywood presents us with a way of accessing a level of reality that is virtual and material, and always in the process of multiple becomings. But does that mean this is a nonnarrative cinema? I attempt to answer this question through a consideration of Ravi Vasudevan's justly famous reading of the deification of heterosexual masculinity in Bimal Roy's 1955 version of *Devdas*.

Although not the most successful version, Bimal Roy's *Devdas* in many ways set the standard not only for cinematic adaptations of Sarat Chandra Chattopadhyay's novel (there have been eight film versions of the novel) but for subsequent filmic representations of romantic love itself. It is often said that Dilip Kumar's performance as Devdas is the epitome of emotive acting (Jaan Production, 2000). The story revolves around the doomed love between Parvati (Suchitra Sen) and Devdas, who are socially mismatched by caste and class, and Devdas's unconscious rejection of the loving devotion of the prostitute-dancer Chandramukhi (Vyjayantimala). As many viewers have informed me, when *Devdas* was released in 1955, new and old conversations concerning the effectivity of the tragic lover, the social constructions of the family and lineage, and women's subordination under Hindu patriarchy all became part of the cultural discourse around the film. We can discern aspects of this in the prerelease publicity for Roy's version. In a full-color "Star Profile" *Filmfare* did with Suchitra Sen, who played Devdas's true love Paro, for the August 5, 1955, issue, we learn that there is a great deal of buzz around Suchitra, "whose real name is Rama Sen," because *Devdas* is her first Hindi film, when she was already "Bengal's reigning box-office queen." Moreover, atypical for Bombay actresses, she is married and the mother of a four-year-old daughter: "In 1947 she married the man with whom, she says, 'I had been in love for as long as I can remember'" (Star Profile, 1955). Suchitra is a thorough professional: "An artiste who takes her work seriously, Suchitra is an early riser and, at the studios, is ready in her make-up well before the others." And she does her own coiffure. The caption reads: "Suchitra has rich, long tresses, and takes great pains over her coiffure. She combs and twines her hair herself." She also drove a "brand new Buick convertible" from Bombay to Calcutta, giving her an aura of modern independence and once again differentiating her from her screen persona. Coupled with the bright-red lipstick in the color photos, these extra-diegetic elements of the star aura both compete with the character Parvati and complete the narrative, because, unlike Paro, Suchitra marries her longtime love.

At a cost of more than $13 million, Sanjay Leela Bhansali's (2003) version was the most expensive film ever made in India. Viewing Roy's *Devdas* today, after Bhansali's recent elaborate and luxurious translation, we can see why Bhansali dedicated his version to Roy: Roy's *Devdas* is a classic of Bombay film making. In that sense, Roy's *Devdas* comes to us from what many call the Golden Age of Hindi-Urdu cinema. For some critics, 1950s cinema was "a transitional period between the popular culture and mixed social audience of the 1930s and 1940s and the mass audience emerging from the 1960s" (Vasudevan, 1995, p. 312). Vasudevan takes issue with such a reading. He suggests that 1950s cinema, in its deployment of "a rhetoric of traditional morality and identity" (p. 314) to bind its imagining of social transformation, already prefigures some of the dominant methods of the subsequent periods of Hindi cinema. More specifically, Vasudevan analyzes how hierarchies of patriarchal power develop around the image of a character. "This character image becomes the authoritative focal point of a scene, occupying a certain privileged position which structures space as a force field of power" (p. 314). In understanding the sacralization of this image, Vasudevan deploys Geeta Kapur's notion of the iconic, where the iconic is "an image into which symbolic meanings converge and in which moreover they achieve stasis" (p. 312). However, in Vasudevan's redeployment of the iconic, the term is used "to situate the articulation of the mythic within painting, theatre and cinema, and could be conceived of as cultural work which seeks to bind a multi-layered dynamic into a unitary image" (p. 312). The crucial point here, as with all effects of closure in film, is that the unitary iconic is *one* effect of the process of decoupage and montage in cinema, not their natural outcome (Guzzetti, 1975).

If this is one effect of the assemblage, let us consider another by looking again at the segment that Vasudevan (1995) argues stages the deification of masculine authority. In Roy's 1955 version of *Devdas*, the first glimpse we get of the adult Devdas is apparently from Paro's perspective. It seems that she is looking at his shoe-clad feet first, with eyes downcast and her head down, in the proper pose of the devoted bhakti-wife awaiting the coming of her Krishna-lover. For Vasudevan, this moment is iconic insofar as the male desire of the camera is established through a process of deification: Dev's feet activate a chain of signs tying him to Krishna, and Parvati to Radha. The film, thus, utilizes "continuity conventions to the highly 'traditional' end of deifying the male as object of desire" (Vasudevan, 1995, p.315). First, for Vasudevan, the ideological effect of male deification is produced through a process of continuity editing. Second, the sequence in question narrates Paro's point of view. Third, the agency of masculine deification is the filmic image. Vasudevan continues: "In anticipation of Devdas's arrival, Parvati hurriedly starts lighting a *diya*, a devotional lamp, and the melody of a kirtan, a traditional devotional song expressing Radha's longing for Krishna, is played" (p. 315). We hear the sound of Devdas's footfall on the stairs—if the filmic image deifies the masculine, it is the ear of Parvarti that opens another kind of agency. Supplemented and split by the soundtrack, Parvati's anxiety to light the lamp before Devdas enters her room is caught by a suspenseful intercutting between her hurriedly lighting a match and shots of the empty doorway. Furthermore,

> [t]he doorframe in this sequence suggests the shrine in which the divine idol is housed. Devdas's entry is shown in a highly deifying way: first his feet are shown in the doorway, followed by a cut to the lighted lamp. Finally his face is revealed. There follows a cut to Parvati, suggesting that this is the order through which she has seen Devdas's arrival. As she looks at him, in a classical point of view arrangement, conch shells, traditional accompaniment to the act of worship, are sounded. The future husband as deity, object of the worshipful gaze, is established by the narration's deployment of Parvati's point of view. (Vasudevan, 1995, pp. 315–316)

Thus, Vasudevan insists that we must "acknowledge the framework of masculine authority within which female desire is finally held" (p. 316). What concerns me in Vasudevan's reading is the excision of zones of potentialities inhabiting this "hold" on Paro's subjectivity, as well as the closed determination of narrative structure by the imperatives of a certain patriarchy. I argue that by considering Paro's agency as a complex articulation of desire, narration, and a field of potentiality, we can also begin to address the overdeterminations that both structure and deterritorialize "the framework of masculine authority."

Thus, if you break this segment down into its constituent frames,[10] you realize that the camera's first cut after Paro lighting the *diya* is an eye level shot of an empty doorway. Rather than simply questioning the actual sequence of editing, this becoming-image allows us to return to the plane of incipiencies and deformations. Is this empty doorway the male under erasure? The shrine of the male idol? Yes, but an obscure, empty doorway is a virtual space as well. It is a potentializing opening: time, space, movement, echo, threshold, passageway, void, potential presence, and means of communication. It is the very site and means of becoming. In that sense the empty doorway is a kind of Deleuzian face that "establishes the prior level of communicability, the 'is it possible?' that precedes the *what* of thinking, saying, feeling ... The face is, therefore, potential" (Rushton, 2002, p. 226). The empty doorway is of the order of potentiality and the virtual. If we can think of this suspenseful doorway as the face of potential, then the elements of Vasudevan's (1995) masculine iconic are organized around and through this virtual passage. We should remember that for Deleuze the virtual intercedes and suffuses the "real as that which is expected, predicated, or even imagined of the real. It is the originary evanescence without the emergence of which nothing in reality could be actualized ... the virtual is the very proceeding of the actual" (Rushton, 2002, p. 226). If the shoe-clad feet of Paro's master are iconic, then no less so is the empty space of the threshold, which is a becoming that emerges from that other, virtual space of the void.

Moreover, if we maintain with Vasudevan (1995) that the camera movement narrates Parvati's perspective, then it in fact narrates a much more complex subjectivity than simply the victim of a male fantasy. Even a becoming-flame of subjectivity itself. Vasudevan misses the full force of the first shot after the iconic feet of Devdas: two unevenly positioned flames signify the devotional *diya* and the dying match that lit it. It is only then that Devdas emerges from the shadows; it is only then that we see Paro not with downcast eyes, but with the direct gaze of an adoring lover. Something other than subjectivity is being signed with these cuts, and with the panning of the camera that precedes them. Indeed, as Paro looks out toward the virtual space of the empty doorway, we can grasp that moment as the repeated assertion of an unequally positioned desire that is not less intense in desire than that of Devdas. Later in the movie she will be made to suffer for it by his hand, as he scars the pure moonlight face of his beloved. However, I suggest that we forget Parvati's indomitable agency at our peril, because another possibility always haunts, indeed enables, the masculine iconic.

It is by keeping in play these other possibilities that the question of whether Mumbai cinema is a nonnarrative form can properly be addressed. What this segment does is open up narrative time to the presentness of the moment. It is through this presentness that anything like suspense is possible in the first place. In his recent thesis on narrativeness, Morson (2003) provocatively asserts that not all narratives have narrativeness, and that the necessary condition for any degree of narrativeness is a process: "The sense of process, the activity of tracing possible futures from a given past, is essential to narrativeness" (p. 61). Moreover, this process must be articulated through a series of presents:

> the present moment must matter. It cannot be a mere derivative of earlier events or dictated by later events, that is, by the structure of the whole ... And what gives a moment presentness? In a phrase, open time. For a present moment to matter, to have real weight, more than one thing must be possible at the next moment. We may define open time as the excess of possibilities over actualities. For a determinist, one and only one thing can happen at any given moment, what did not happen could not have happened. In open time, at least one thing that did not happen could have. (pp. 61–62)

Narrativeness, then, is constituted by process, by presentness, by events that exceed their actualization, by contingency, messiness, and unpredictability. And, of course, suspense (Morson, 2003).

It is in this sense of narrativeness that Hindi cinema at its most challenging can help us to rethink the very nature of what constitutes narrative itself. The empty doorway, the virtual space of the void, not only frames the iconicity of Devdas but also launches that other line of flight that describes Paro's desire. A nonactualized event, it is a present moment because more than one thing could and does happen through its narrative processes. That is also its suspense. The framework of masculine authority and feminine agency staged through this segment, therefore, is an effect of a process that allows us to consider other potentialities[11] that not only haunt but radically enable the iconic narrative of male sacralization.

This potential Otherness constitutive of the image-sound regime diagrammed above is both captured as algorithm and potentialized as a becoming-flame in one segment of Bhansali's *Devdas* (2003). The segment is from the song "Silsila yeh chaahat ka" (An Affair of Desire) and is an obvious quotation of precisely the scene that Vasudevan (1995) focuses on. It is staged in part as a devotional prayer to the goddess Durga (an incarnation of Devi or Parvati), who in the last shot of the segment is shown in the traditional frontal pose before a lion, in red sari, her eight arms wielding various weapons ready to wipe out all the evils of the world. The lyrics, however, tell of an affair whose flame has never been extinguished, and that is the visual conceit of the song. A whirling Parvati (Aishwayra Rai), dressed as Durga, defies the mockery of her girlfriends and cradles a *diya* in the palm of her hand, keeping the flame burning brightly through rain, wind, and pirouettes. Bhansali deploys an active camera, with numerous pans and angles, continuous 360-degree shots, and rhythmic movement from one color-space to another for the sake of a different deification. The perfusion of color from lyric sequence to musical refrain is dominated by the red of Durga's sari and the vermillion that explodes in bursts of reddish haze. In the initial sequence this montage-sound-color strategy builds with the expectant tempo of the lilting song sung by Shreya Ghosal, carefully cropping Parvati's face from visibility. This movement reminds us how the faces of gods and goddesses in Hindu temples are often hidden before devotion begins; thus the *darshan* circuit, the gazing on and the gaze of the goddess, is activated through devotion.

Until a sudden illumination courtesy of digital animation, Parvati (Ghosal-Ashwarya) starts to sing ("Silsila yeh chaahat ka na maine bujhne diya") as a flash of lightning illuminates her face. She lowers her arm, and the flame "catches the mirror" on her *bindi* and, digitally enhanced, momentarily flickers, like a moment of nonmotivated cinematic excess (Thompson, 1986). The potentializing face here is pixilated, and these pixels are massaged in a variety of ways through algorithms common in commercial digital filters: the face today is an open doorway that confronts the viewer as a graphical user interface. Together with the signature color saturations and active camera that characterize Bhansali's style, the digitized face engulfs the viewer in a careful orchestration of movement, sound, color, so that Parvati's rhythm turns into flame (Black, 2005; Ghosh, 2005) This is a qualitatively new sensory-motor schema that implicates the body of the viewer through a particular synesthesia of sound, light, texture-color, duration, movement. Bhansali's quotation is in fact a refunctioning of the entire assemblage that enabled Roy's "deification of masculine authority," staging cinema as an event of a new deification, new relations of power, establishing the connectivities of a new assemblage. These assemblages are not comparable in the sense that the (post)modern is the maximization of the dream of the Other. The embodiment of the senses, the affective multiplicities, are qualitatively different in both. The nature of this difference is what is at stake here. The excesses of Bollywood are not the purple pleasures of a cinema expanding in all directions but the contagion of events opening narrative to the virtual futures constituting it.

Notes

1. *Mother India* (1958, dir. Mehboob Khan) starring Nargis and Sunil Dutt was among the most important articulations of the national in the idioms of the cinematic epic. It won national awards and loud acclaim from public figures, including Nehru. In their roles as national representatives in the mid-1950s, Raj Kapoor and Nargis toured the Soviet Union and were received like royalty by adoring Russian fans. See Bakshi (1998).
2. See "Objectives of film certification" on the Central Board of Film Certification Web site (http://www.cbfcindia.tn.nic.in/guidelinespage2.htm): (a) the medium of film remains responsible and sensitive to the values and standards of society; (b) artistic expression and creative freedom are not unduly curbed; (c) certification is responsible to social changes; (d) the medium of film provides clean and healthy entertainment; and (e) as far as possible, the film is of aesthetic value and cinematically of a good standard. See also Noronha (2004).
3. The latter three critics have been central in my reimagining of social practice across a variety of vector fields. I have never worked specifically with Partners for Urban Knowledge Action Research (PUKAR), but the work of Phadke, Sardesai, and Shrivastav both with that organization and as autonomous artists (visual art and novels) and scholars (the question of security and gender in urban space, the politics of entertainment-

shopping space, the creation of literary publics across linguistic boundaries, the strategies of editing and administrating, and the reforming of literary genres ... but above all a deep love of finding the thresholds of thought in a profound engagement with the world and its sensorial) has left a deep impact on my thought in these pages.

4. DCT, short for discrete cosine transform, is a technique for representing waveform data as a weighted sum of cosines and results in what is called "lossy compression," which approximates some of the coefficients to reduce the amount of data.

5. A proprioceptor is defined as "[a]ny sensory structure which receives stimuli arising within the tissues (other, usually, than the viscera); esp. one concerned with the sense of position and movement of a part of the body" (*Oxford English Dictionary*). Hence, proprioception is the reception of information by proprioceptors and its interpretation. The context for proprioception is in fact neural pathways, as well as the memory of the flesh—in muscles and fibers. It is the stuff of both neuroscience and science fiction. To control and modulate proprioception directly seems like the very dream of not only Western (military) science but advertising as well (Hollywood and the military coincide in the affective economies of societies of control, as Patricia Clough has pointed out to me). For two interesting recent science fiction stories dealing with proprioception, see Doctorow (2005) and Ford (2006).

6. For "biogram" see Brian Massumi (2001): "It has been suggested that extending the concept of the diagram into the biogram might be a vector worth pursuing. Formal topologies are not enough. The biogram is a lived topological event. It is onto-topological. It is the event of experience folding back on itself for its own furtherance, its continuing becoming. Onto-topological means ontogenetic. The biogram is experience reaccessing its powers of emergence, for more effect" (p. 207). Conceptually, there are direct precursors within Indian film and media criticism that have enabled this thought of biogramming media assemblages. There is the line of analysis that has shown how Mumbai cinema's fragmented mode of production can be diagrammed (exemplary here is the careful work of both Ravi Vasudevan and Madhav Prasad). There is also the work on the hybrid media genealogies of the popular film form; as Esha Niyogi De (2007) remarks, "Spearheaded by the cosmopolitan Parsis who had acquired wealth and prominence through modern business enterprises and western education, this theater was an important channel through which translocal influences (Victorian melodrama, Shakespearean and seventeenth-century British drama, non-western oral traditions such as Persian folktales) flowed into Indian performance traditions in the late nineteenth and early twentieth centuries. Although the dramatic tradition had originated in the colonial academy, its producers and performers soon undertook commercial ventures, becoming attentive to popular

tastes ... these alterations involved changing the very ways in which the 'constants' of human existence were being conceptualized in performance" (p. 27). My argument is that this reconceptualization of human existence happened through a particular shift in the nature and connectivity within and amongst these media: a new sensorium came into being with the birth of popular mass culture in twentieth-century India. This is what I am calling the media assemblage.

7. What is a *hat ke* fold? *Hat ke* means something which is a little awry from the norm, a kind of masala film whose effects are not trackable by the functional segmentations of a population. Something tangential, but always also at only one remove. A fold? When the fetish of white virginal skin (Bollywood's heroines, from Saira Banu to Kareena Kapoor) folds into the fetish of the good South Asian (the Hindu nonresident Indian computer techie), a new potential multiplicity is in the becoming.

8. Traversing these sites like a dream/nightmare, the touch of cinema is a mode of contagion. And sometimes vision itself becomes tactile: Hindi cinema is, in other words, a haptic medium. In thinking about commercial Hindi cinema as a haptic media event, I am following the work of both Deleuze and, especially, Laura Marks (1999). As Daniel Smith explains, in his elaboration of Francis Bacon's *Logic of Sensation*, "what Deleuze ... terms haptic space (from the Greek verb *apto*, to touch) is a space in which there is no longer a hand-eye subordination in either direction. It implies a type of seeing distinct from the optical, a close-up viewing in which 'the sense of sight behaves just like the sense of touch'" (p. xxxii). Indeed, Deleuze, in a way that suggests something of the "to come" of the future anterior, warns that "[o]ne can remain entangled in the figurative givens and the optical organization of representation; but one can also spoil the diagram, botch it, so overload it that it is rendered inoperative ... The diagram is thus the operative set of asignifying and nonrepresentative lines and zones, line-strokes and color patches. And the operation of the diagram, its function, says Bacon, is to be 'suggestive'... Because they are destined to give us the Figure, it is all the more important for the traits and color-patches to break with figuration ... In order to be converted into a fact, in order to evolve into a Figure, they must be reinjected into the visual whole; but it is precisely through the action of these marks that the visual whole will cease to be an optical organization; it will give the eye another power, as well as an object that will no longer be figurative" (pp. 82–83). Precisely: the affect of touch. Extending and transforming this experience of vision through our memory of other senses (touch being only one of them), Laura Marks argues that the skin of a film "offers a metaphor to emphasize the way film signifies through its materiality, through a contact between perceiver and object represented ... to think of film as a skin acknowledges the effect of a work's circulation among different audiences, all of which mark it with

their presence ... film (and video) may be thought of as impressionable and conductive, like skin ... I want to emphasize the tactile and contagious quality of cinema as something we viewers brush up against like another body" (p. xii). Rather than reducing the haptic quality of media assemblages to the function of metaphor, what we see is that the haptic opens the habitual pedagogies of cinema to its potential contagions, constantly transforming its own conditions and contexts of operation through interpenetrating multiplicities: satellite TV, fashion-able desires, the latest saris, instantaneous Internet, *Stardust*, fan identity, affect, Shahrukh's aura, the slow duration of a movie set. Methodologically, the diagram of the media assemblage would begin from the point of exhibition as event; from the concept of audience as social flows desiring; from the positionality of a criticism that is intent upon its own functionalities (or nature).

9. From the perspective of a media assemblage, we must note that *Kaante* was one of the first films released with explicit tie-ins with the multiplex boom that was just getting under way in India. Its theatrical release—and it was simultaneously released in Los Angeles, New York, Mumbai, Delhi, London, and Kolkatta—in India was tied in with specific marketing schemes at various multiplexes. For instance, at INOX Baroda the entire staff grew goatees, dressed in black, and set up a huge vault that patrons were invited to crack. See Ramakrishna (2003).
10. We should note here that this criticism is enabled, but not enclosed, by a certain advance in technology. DVD-video capture technology enabled me to go backward and forward frame by frame to analyze the exact configuration of decoupage and montage. However, these signs of another agency had already been read by Vasudevan in his recognition of the suspenseful empty doorway.
11. What are some of these "other potentialities"? For instance, we can consider that Paro's subsequent proud rejection of Devdas is made possible by the complexly delineated agency given by this segment, or that at a particular moment, Devdas could very well have run off with Paro, if only, if only ...

CHAPTER SIX

Situating the New Media: Reformulating the Dalit Question

P. Thirumal

THIS CHAPTER POSES a series of interrelated questions about technology, development, and democracy by reconstructing an Internet campaign connected to the public controversy that surrounded the alleged beating of a Dalit employee in a reputed Non-Governmental Organization (NGO) in India during 2003. The term "Dalit" refers to the untouchable Hindu caste, and it is increasingly being used as a political category (Guru, 2004, p. 257). Its legal equivalent is the constitutionally designated term "Scheduled Caste."[1] The content of the communication campaign initially included the alleged (institutional) violence committed against a socially powerless individual. But it grew to envelop other areas, including the dominant rhetoric of development in the NGO sector. It is equally necessary to reflect retrospectively on the channel of communication and the language chosen by the campaigners—the formidable Internet and the hegemonic English language.

The controversy sprang from an anonymous email, and the campaign was sustained through a radical political organization. The campaign managed to force an investigation from well-known human rights activists. But their scathing report, which revealed the nondemocratic character of the NGO in question, was treated with scant respect.[2]

The NGO in question is Deccan Development Society (DDS), a large rural development organization working in the semi-arid tropic region of Medak district, Andhra Pradesh, in India. Though email participants in the campaign belonged to different organizations and various political persuasions, this chapter confines itself to three important actors: Spandana, an anonymous e-group that is purportedly associated with the origin of the controversy; "Kula Nirmoolana Porata Samiti (KNPS),[3] a Dalit wing of the radical Left, which launched a campaign in relation to the controversy"; and DDS itself, the perpetrator of the alleged incident.

The Internet, the English Language, and the Dalit Citizens Network: Technology and Ontology

The controversy discussed here became public because the whole campaign was conducted on the Internet. It is useful to make a distinction between an Internet public and the public that is constituted in and through traditional media such as print, satellite, and cable television. The term "interactive" characterizes the new media vis-à-vis older technological media. Whereas the older technological media failed to represent Dalits in any meaningful way, either as producers or as consumers, the new media allow the Dalits to imagine themselves as producers of messages and images. It is interesting to note that the Internet is providing a platform for the educated Dalits to participate in issues concerning their community on a par with other dominant groups.

Radio, television, cinema, and newspapers have been extremely inaccessible media for Dalit issues and Dalit participation (Guru, 2001). There is a paucity of academic research to indicate that traditional media in India played a negative role in the construction of the Dalit question (Jeffrey, 2000; Prasad, 2004). Pre-Independence mainstream print media mostly bypassed the Dalits. However, publications emerged in the first few decades of the twentieth century, as a limited Dalit public sphere emerged. Ravi Kumar, the Dalit intellectual and cultural theorist from Tamil Nadu, has written (Anand, 2005) about the presence of a nominative Dalit press in the early twentieth century in Madras (now Chennai). Bhoite (cited in Wagle 1999, pp. 31–41) points to journals published by activists during the vibrant liberation movement in the state of Maharashtra. Of special note is the work of Bhimrao "Babasaheb" Ambedkar, a leading Dalit scholar and anti-caste reformer who rose from his untouchable birth to become an architect of India's Constitution and a member of the first post-Independence parliament. In 1920, he launched *Mooknayak* ("Leader of the Silent"), a journal that criticized the lack of political will among the untouchables to advance their status. (See, for example, Buddhiwadi Foundation.) Other outstanding Dalit individuals early on took up the task of cultural production to attack Hindu nationalism and colonialism, including Iyothee Thass in the south and Swami Achhutananda in the north, both of whom launched newspapers of importance to the Dalit community from the late nineteenth century into the early twentieth century. However, these efforts were not comparable to the near-absolute monopoly of the press by the Brahmins. In 1945, Ambedkar wrote:

> The Untouchables have no press. The Congress Press is closed to them and is determined not to give them the slightest publicity. They cannot have their own press and for obvious reasons.... The staff of the Associated Press of India, which is the main news distributing agency in India, is entirely drawn from Madras Brahmins—indeed the whole of the press in India is in their hands and they, for well known reasons, are entirely pro-Congress and will not allow any news hostile to the Congress to get publicity. These are reasons beyond the control of the Untouchables. (cited in Anand, 2005, p. 172)

Jeffrey (2003) makes a similar point:

> The fact that no Dalit men or women worked in minor editorial jobs on Indian-language dailies meant that aspects of the lives of Dalits were neglected. And the fact that no sizeable daily in India was owned or edited by Dalits meant that stories about them were unlikely to receive the constant, sympathetic coverage of stories about, for example, the urban, consuming middle class. (p. 178)

Unlike the West where economic and social democracy precedes political democracy, in India there has been a reversal of sorts. The post-Mandal situation in the early 1990s created favorable conditions for the issue of representation to be brought into sharp focus.[4] A brief detour on Mandal here would help bring out sufficiently the role that Mandal Commission, though thwarted in its implementation, served in facilitating discussions regarding Dalit issues in the ensuing years. The surfacing of spaces such as www.ambedkar.org and www.dalitstan.org in shaping up the emergent Dalit identity is not an unconnected formation in a prototype that slowly makes its entry into the world of media, making it lose its character of being a mainstream enterprise.

The implementation of Reservation by the former Prime Minister V. P. Singh in the late 1980s for the Backward Classes as articulated in the Mandal Commission Report (B. P. Mandal was the Chairman of the Committee constituted by the Government of India in the 1970s) represented a move towards social democracy, which was stridently resisted by the upper caste urban population. In the current times, both Hindu nationalist politics and the avowed acceptance of free market by the Indian State promote rather than expectedly discourage identitarian issues so much so that issues of identity have become very legitimate questions in the public sphere. The flowering of literature on Ambedkar and Dalit studies in universities and higher educational institutions should be seen in this newly forming context.

James Curran states that "a central role of the media should be defined as assisting the equitable negotiation or arbitration of competing interests through democratic process" (Curran, 1991, p. 30). In the light of Curran's statement, it could be argued that the non-upper castes were involved in elaborating the idea of radical democratic practices in connection to the media. There has been a demand from the Dalits and the Backward Castes to reach an evenhanded negotiation or settlement of contending interests through the democratic mechanism facilitated by participation in the media. A few active Dalit Internet groups, such as the aforementioned www.ambedkar.org and www.dalitstan.org, continuously seek and shape the issues that are crucial in the construction of a Dalit history as well as identity through the democratic participatory forums facilitated by the new age media of the Internet. However, this is not to overstate the case, or claim that the Internet has emerged as a viable alternative to mainstream media networks. The Internet, in some limited sense, seems to provide a locus for writing the lives of the nascent Dalit middle class in semantic if not syntactic English.[5]

The campaign, discussed in connection to the controversy introduced in the beginning of the chapter, was carried out in the English language. Both English and the Internet are in some sense elite. They exclude any subaltern participation. However, the argument of this chapter is the contrary. The male Dalit bourgeoisie, who have strayed into the English language and possess minimal competencies and cultural capital, though referred to as "stutterers" and "mutterers,"[6] are subverting the link between the political and the local. In the classical liberal sense, "political" stands for the institutional authority of the state; "the local" in a neoliberal sense stands for insulation from central authority.[7] The state is supposedly not the author of the local. However, in the age of globalization and unfettered market relations, the local cannot be held in its place.

The cohabitation of the English language with the post-Indian Nehruvian state has enabled the upper caste to emerge as the producer and custodian of a syncretic national culture. The halting and gradual appropriation of the English language by the lower castes in the post-Mandal period threatens this artifice and the upper caste's notion of India, even as the accumulation of symbolic and cultural capital has been a struggle for the disadvantaged castes.

The question of how to root the Dalits, or write about the Dalits, is an intriguing one. The Constitution of India (Basu, 1987) is considered a virtuous document by most educated Dalits because the category called Scheduled Castes and Scheduled Tribes is prescribed as a legal entity requiring special treatment. This foresight is attributed to the Dalits' venerable leader Dr. Ambedkar.[8] Inscribing a large pan-Indian category of institutionally and historically disadvantaged, heterogeneous groups as potential citizens is of revolutionary import. It is through writing that such a category is formulated. For the first time, a history of a radically new people is constituted almost without their knowledge—almost without their knowledge because it is only in and through writing that they can claim this history. The inscription of Scheduled Castes in the Constitution presumes a relationship between writing and the ontology of the Dalits. Though debates in the Constituent Assembly preceded the writing and founding of the Indian Constitution, it is essentially a written document. Strangely, this document seems to have been informed by the history of the future.

Today, the newly constituted people claim this history and writing, and this claim appears to subvert the expected chronological progression of events. The linear world's historical present has to extend itself forward and not backward to accommodate this claim.[9] The efforts of Mayawathi, the former Dalit chief minister of Uttar Pradesh, to establish an Ambedkar park replete with statues of subaltern icons represented nothing other than an act of setting fire to a received national history. This is not to deny that Dalits being ushered into time/ history through inscription in the Constitution was actually a precondition for this explosive reconstitution and rewriting. This is certainly not a silent revolution, but it is also not a radically egalitarian project.[10] The mapping of this populace as Scheduled Castes in the Constitution and the redefinition of the Dalits post-Mandal reflect their transition from being the constituents of an administrative category initially to being rewritten as a politically volatile category (Guru, 2004).

It is important to mention that unlike the Scheduled Tribes in India, Dalits historically have not been a territorial community; they are largely located in rural India. Theoretically, the technology of the Internet provides an opportunity for the Dalit male bourgeoisie inscribed in—and inscribing—the increasingly globalized national script to deterritorialize the massive population of the Dalits located in the rural-local. In another context, Derrida, in conversation with Stiegler on the nature of the addressee in print culture and new media, remarked:

> The addressee has never simply been a passive receiver. If we recall, as you just have, that access to writing in the classical sense was the condition of citizenship, this is the very thing that is changing today. The question of democracy, such as it has been presenting itself to us here, may no longer be tied to that of citizenship. If, that is, politics is defined,

> as up to now it has been, by inscription in a place, within a territory—given, lost or promised. All the problems we have been talking about with reference to a technology that displaces place: the border is no longer the border, images are coming and going through customs, the link between the political and the local, to political, is as it were dislocated. (Derrida & Stiegler, 2002, pp. 56–57)

Since the Dalits are not inscribed in a particular place or territory, they constitute perhaps the only community other than the Brahmins to display an eagerness to share a pan-Indian identity. This eagerness can be realized through the use of and familiarity with the new media by the nascent Dalit middle class.

The use of not particularly syntactic but semantically comprehensible English opens up a symbolic universe populated with liberal notions of citizenship and democracy. The Internet does not confine itself to those who are well-versant in the conventions of literary and journalistic English. The minimal presence of standard or conventional, and literary and official English language among the newly educated class of Dalits leaves them with only a rudimentary capacity to make any distinction or erase the distinction, if it exists, between the figurative and literal nature of their modern expressions and rendititons.[11] These people, who lack confidence in the realm of conventional modes and mediums of writing, seem to log on to the world of the Internet without undue embarrassment. It is in this sense that the democratic character of the Internet opens it up for subaltern appropriation.[12]

In received histories of the West, print democratized the written word through a system of school education in Western society.[13] The print represented the first industrial commodity and the first capitalist enterprise in the West. According to Benedict Anderson (1983), print capitalism played a constitutive role in the arrival of nation-states in the West. It gave rise to his famous thesis that nations are imagined communities. But, unlike in the West, if one traces print's trajectory in India, we see that print and capitalism did not clearly go together. In India, colonially mediated print and colonially mediated capitalism did not make the connection they seem to have made in the West. Though print capitalism was not a primary definer of Indian nationalism or nation, it was still the basis for providing a cohesion that did not exist before colonialism. In fact, it is only in the post-Independence India, and that too during the past three decades, that capitalism has made a foray into the print industry in India (Jeffrey, 2000). Print capitalism, essentially, is a post-Independence phenomenon in India.

It is not surprising that in the pre-Independence period, print bypassed the Dalits for the simple reason that there were few educated Dalits. In the post-Independence period, despite the gradual entry of the Dalits into the educated class, the situation has not changed drastically. The absence of Dalit representation in the print media in terms of ownership stake in newspapers, as newsmakers, accounting for newspaper content, and constituting readership, continues to mark this period.[14] The reasons that rendered Dalits invisible in print media include the lack of social and cultural capital and the lack of literary skills among the Dalits, leaving the educated Dalits with the onus to build networks based on print culture.

Among other things, the lack of the print culture or an unfavorable disposition towards it, due to the unfamiliarity engendered by the denial of access to it, contributed to the lack of cohesion in the national Dalit community. Although a national Dalit community existed insofar as Scheduled Castes are recognized by the Indian Constitution, this legal/administrative entity has not been adequately transformed into a cultural entity. Only when this administrative entity becomes conscious of itself can it evolve into a political entity with a cultural memory.[15] Thus exclusion from print media has had far reaching repercussions on the Dalit identity and history, as it did not help build a cohesive national Dalit community or inscribe a legitimate Dalit historiography.

The contemporary socioeconomic changes brought about by globalization raise many issues for the Dalits and other marginalized people. The most commonly heard criticism from the Left is that globalization, in its nexus with global capital, will fling the already helpless Dalits into further deprivation. While the Marxists make an economic argument, there are positions that speak about the corresponding cultural alienation wrought by globalization. While print seems to offer itself as the medium of modernity, with its connections—tenuous or strong—with capitalism and nation-states, the Internet as a medium characterizes globalization. While national capital through centralized technologies produced an administrative category, global capital through the deterritorialized medium of the Internet seems to have the potential to translate the administrative entity into a cultural category. For the Dalits, the administrative category provides conditions for negotiation, whereas any culturally embedded practices appear exclusionary. Paradoxically, it is the Internet that offers the newly emerging educated class of Dalits a space to imagine themselves as a pan-Indian community.

The Rhetoric of Development

Having ruminated on the relationship between technology and an ontology of the Dalits, between writing and the primary absence of the Dalits in the written space, I proceed to examine an alleged incident in which a Dalit was slapped on the face with a slipper by an upper-caste executive of the Deccan Development Society (DDS). It is only in language that this alleged act becomes an event. The face on its own cannot reveal the act. Perhaps the victim failed to perceive the act as humiliation, insult, or indignity. It is the argument of this chapter that the alleged act has the potential to be constituted as a violent act only through an interrogation of the dominant rhetoric of development prevalent in the NGO sector. The chapter also looks critically into the category *community*, especially the idea of "local community" posited by multinational NGOs to locate the Dalits. In contrast, I understand the Dalit community as having always inhabited a national identity, both spatially and historically.

Before examining the rhetoric, it is important to narrate the event. In a linear sense, the story has five scenes. The first scene consists of the anonymous email, making the allegation against the chief executive of DDS. Next, DDS denies the allegation. The third scene consists of KNPS demanding, through an Internet campaign, public scrutiny of the allegation. The fourth scene relates to DDS agreeing reluctantly to face a fact-finding committee. The fifth scene is a final rebuttal to the indictment of the fact-finding committee by the members of the board of directors of DDS.

For Spandana, the issue involved gross misconduct of the alleged aggressor toward his Dalit driver.[16] Apart from sounding the allegation, the anonymous email pointed out larger issues such as the unregulated nature of the development sector. This incident opens up a debate on development, on how to read the margins, and on what constitutes the margins. In academic discourse, the two dominant views regarding development are the liberal-national-modernity perspective and the communitarian perspective, represented by the figures of Nehru and Gandhi, respectively. The communitarian perspective is also referred to as antidevelopment perspective. These two positions are pitted against each other. The communitarians have argued that the Nehruvian liberal state stood for technocratic governance. In the course of time, such a form of governance appears to have paved the way for the oppressive homogenization of society. The indifference of the technocratic state to the perceived differences and plurality of society further led to the emergence of identity movements such as those of language, region, caste, religion, and gender. The national community presumably disintegrated into these various fragments. In fact, the emergence of identity movements coincides with the emergence of the NGO sector in the late 1970s, which is taken to symbolize popular dissatisfaction with the instrumental nature of the Nehruvian paradigm of development. The NGO sector addressed the issue of engendering better life chances for rural communities without disturbing their existing associational patterns. This seems to be the dominant NGO idea of development.

This discursive frame suggests that the local is good and should be insulated from the national/ global to maintain organic continuity and to perpetuate the same for posterity. DDS could be contextualized within such a discursive frame. In another context, elaborating the discursive frame of civil society organizations, Michael Watts (2004) observes:

> Much of what now passes for grassroots anti-development initiatives, or the new sorts of anti-system or anti-globalization movements, would fall within the circumference of this more polemical notion of community. In much of this work there is a tendency to read community as an unalloyed "good": to highlight, to the exclusion of all else, the purportedly ethical, moral and social virtues of the community. The "warmly persuasive" qualities of community have an echo in the related notion of civil society—understood as a complex and dynamic ensemble of legally protected, self-organizing and self-reflexive non-government organizations in tension with one another and the state—which is often seen as a realm of freedom, pluralism, participation and reflexivity. Civil society—and community—suffers from a sort of Manicheanism that afflicts the state with coercion and society with freedom, as "good living, warm and whole." But, as John Keane notes of civil society, the community is 'plagued by endogenous sources of incivility'...
>
> But what is incontestable is the multiplicity of forms and historical circumstances under which community reappears as a basis for participation in public affairs, civil society and in the circuits of capital. At this historical moment we are awash in communities, and the "self-governing community" is one of the defining articulations of neo-liberal rule. Rose suggests that the community resurgence of the 1980s and 1990s in the trans-Atlantic economies—from community policing to community partnerships—cannot be grasped outside of the collapse of socialism and the so-called neoliberal grandslam. Here a new generation of communities arise from the ashes of state withdrawal and speak the name of civic renewal (Putnam is, of course, its avatar). Whether and how such communities can become something subversive—what Foucault called the movements of subjugated knowledges and their resurrections—depends in part on how deeply the state and capital have penetrated the trenches of civil society. (paras. 6 and 7)

In such a framework, the written medium appears redundant and superfluous. In fact, the written medium is suspected of centralizing and tacitly marginalizing the pristine oral rural setting. In the place of this redundancy of writing, DDS offers rudimentary radio production and videographic skills as enabling technologies for the illiterate Dalit women of Pastapur.[17] While there may be several other enterprises associated with this NGO and so many other ways of talking about this entity, this chapter seeks to retrieve and engage with the public controversy mentioned above.

My main objective is to explore ways of reading the discourses on the alleged violence. It is very unlikely that the illiterate Dalits of Pastapur ever participated in this public debate, given the inaccessibility of Internet technology and the fact that the campaign was conducted in English. It was Spandana that initially brought to light the controversy by raising several issues related to undemocratic managerial practices in the development sector. The email, accessible originally to only the NGO sector, was inadvertently leaked out to the non-NGO world.[18] The debate went public after the intervention of KNPS, which insisted on legal/ public scrutiny of the incident and also invited responses from Marxist groups, Dalit and human rights activists, progressive academics, and intellectuals.[19] While Marxist academia and activists are yet to come to terms with caste as a powerful analytical and experiential category, many radical Marxist groups have started engaging with issues relating to caste. KNPS seems to be a product of such a historical experience. In a neoliberal age in which plurality is presented as an antidote to the onward march of homogenizing global capital, where egalitarianism has come to have purely antiquarian value, KNPS appears to be burdened with these competing themes of differences and egalitarianism.

After persistent Internet campaigning by KNPS for a full-fledged inquiry into the incident, DDS agreed to face a public enquiry. The fact-finding committee examined the case and came down heavily on the undemocratic manner in which DDS functioned. Reacting furiously to the comments made against the DDS board of directors, one of the senior members of the board responded:

> As an activist in the women's movement, I am very aware of the issues related to identity politics. However, I can state that every issue and difference between a man and a woman is not necessarily a gender issue. In the same way, I feel that every issue between a Dalit and a non-Dalit is not a Dalit issue. I do wish that the committee members had stated their bias in their report so that we can also judge the case and your findings in an impartial way.[20]

This mail captures the tension between the two positions—imaging the Dalit as a communitarian subject in some situations and a noncommunitarian subject in other situations. When a man beats a woman, it should necessarily be read as gender violence; if an upper-caste person beats a Dalit, it should not necessarily be read as caste-related violence.

In this case, why should DDS describe itself as an organization that has "emerged as [a] Dalit women centered voluntary organization over the decades"?[21] It is necessary to unravel this claim and its contradictory consequences for the very same proposition in its chartered motto. This statement amounts to saying that DDS has provided congenial conditions for the production of powerful Dalit women. If this is the case—that is, if the Dalit women are so empowered, and, as a corollary, all Dalits—it is not clear why DDS has continued to operate in this area for more than two decades. If caste is not the locus of authority in backward rural Andhra Pradesh, what other bases of authority operate in large, corporately funded NGOs in the country? I participated in this campaign and problematized the issue in an email to KNPS:

> Why Dalits in Pastapur are supposedly silent while their counterparts in the cities are agitated over this (alleged) violence? Perhaps, it has to do with the idea of Dalits perceiving themselves as a local community belonging to the DDS Empire at one end and the national empire at the other end.
>
> A few years ago, DDS released a calendar entitled "Local Communities" or something approaching that; they have programmes for retrieving local knowledges of the toiling classes and they have persuasively argued with the National Government for setting up a "Community Radio" station for poor women—to be read as Dalit women. It is said that both the international media and the national media have always been at the beck and call of DDS for publicizing their programmes.
>
> The singular ideological violence that NGOs can bestow on the Dalits is by describing them as grounded, rural and local communities. While the Brahmins heading these institutions are already a global community, the effort to reduce the Dalit national community to local histories and knowledges is revealing. It is equally absurd for some NGOs to proclaim the Dalit community as a global entity with enormous reach and power.
>
> At a time when the Nehruvian liberal state lies decimated and the neo-liberal paradigm reigns supreme, the intervention of NGOs through the philosophy of "small is beautiful" seems very enticing. The Nehruvian liberal state created the Dalit national community, the Neo-liberal

> regime has fragmented this community through the effort of the NGOs to a level where their bargaining capacity either with the state, the civil society or the market has dipped phenomenally. In fact, the Nehruvian State rhetorically spoke of inclusion but the current regime refuses to even rhetorically make such a submission.
>
> It has been argued that the failure of the Nehruvian state to ensure citizenship entitlements led to the reasoning that communities had to be empowered instead of being assigned a socially vacuous citizenry. In other words, communities were to become the central actors in place of citizens. It was precisely at this juncture that the arrival of identity politics coincided with the growth of NGOs. While the identity politics led to the assertion of backwardness among communities, the NGOs stealthily de-politicized the development sector. The NGOs were translating the Dalits from a modern administrative category to a pre-modern cultural category.
>
> The administrative category both facilitates and inhibits the autonomous growth of the pan-Indian Dalit community. On the other hand, the idea of them being described as "local community" only regulates their already regulated consumption patterns and legitimate aspirations to enter modern spaces.
>
> It is common knowledge that local customs and rituals produce local communities. Perhaps, "padapuja" is very central to the production of the "fabled local community" at Pastapur. Balaram, the (supposedly) adopted Dalit son of Sateesh can only aspire to be a driver. One of the singular contributions of DDS seems to be the building of a deep awareness of this ascribed position where "pada pooja" is expected to be accepted in its stride by the Pastapur Dalit community. Sadly, such awareness is lacking amongst the urban Dalits and therefore this anger...[22]

This passage explores the cultural resources that DDS powerfully employs to reimagine Dalit ideas and experience. It puts forward the idea that the commandment is to love thy neighbor in a practical sense and not in a dis-empowering sense. Perhaps there is a need for the Dalits radically to disinherit local communities such as Pastapur and stake a claim on the idea of nation. The distinction between rural and urban appears diversionary and dubiously imposed in this information age. India need not live in its villages. Caste is an enduring reality and it has no spatial preferences. But the Dalits need to make a choice, a spatial and a historical one.

The Dalits and Democracy

One of the challenges that modern India offers to intellectuals and laypeople alike is an opportunity to broaden and deepen the democratic character of the country's society and polity. It is in this context that constructing political identities, especially the citizen identity, assumes significant import and becomes an urgent task. Notwithstanding the clash of competing identities and their empowering and disempowering implications for a vibrant and plural democracy, this chapter assumes that any emancipatory discourse related to the pan-national Dalit community must situate the community on the borders of both communitarian and citizenship identities. I offer a minimalist definition of a citizen. The citizen is considered a member of a national community. The national community is conceived of as a horizontal community.

In recent times, there appears to have been a persistent and justifiable demand for the invocation of community rights alongside citizenship rights. With globalization, there is a general apprehension that the nation-state is on a downward slide, taking the national citizen along. While it is difficult to equate the ascent of the Bharatiya Janata Party with the descent of the nation-state, it is possible to associate the BJP's ascent with the erosion of the national citizen. The fall of Yugoslavia and the general anarchic community upsurges in different parts of the world have called for a more nuanced understanding of the categories "community" and "citizen." The citizen has become an impoverished concept, and the national space is seen as oppressively homogeneous with only a hollow enactment of the rituals of democracy.

In this context, it might sound out of place to argue for reinstating the "national citizen" as an analytical and political category. However, I argue for precisely that, because such a move seems more appropriate in terms of re-situating extremely vulnerable pan-national communities, such as the Dalits in India, to positively engage with the culture of managerialism and neoliberal attitudes. Many political leaders seem to cloak themselves as managers rather than political representatives, like the former chief minister of Andhra Pradesh Chandra Babu Naidu, who called himself the CEO of his state.

From a theoretical perspective, the communitarians seem to suggest that they are positing an alternative modernity, a modernity that provides a corrective to the modernity derived from the Western experiences of the Enlightenment. The communitarian position argues for containing the onward march of global capital, which is perceived to be responsible for disintegrating organic premodern communities. An extreme rendering of this stance has held that poverty, along with its corollary development, is only a discourse and lacks any tangible implications for the masses. In fact, development itself is seen as a concept that is sold to the non-Western world to perpetuate the West's domination. The communitarians perceive themselves as directly opposing the national modernity represented in the Nehruvian liberal state. This liberal national modernity thinks of itself as having the potential to create national citizens out of individuals frozen in anachronistic communities. Capital and development occupy a central place in the unfolding of this modernity.

This chapter points out the significance of both the communitarian and national modernity traditions of thinking for the Dalit life world. As Sharmila (Rege, 2000), in her exposition of the Dalit feminist standpoint, observes, "questions that are posed as 'to be for' or 'against' modernity underrate the dialectics of modernity and development and overlook the possibilities of reflexive development" (p. 493).[23] The Dalit feminist standpoint, rooted in the reflexive modernity of Phule and Ambedkar, seeks to develop such possibilities. This is a third possibility, an alternative to the communitarian and the national modernity paradigms.

One of the scholarly stories of the nation runs thus: first, colonial capital disintegrated the essential premodern community and in its place posited the illegitimate national community; second, the national capital, through its rhetoric of imitative development, displaced the national community; finally, the last resistance of local communities, through their reserves of local knowledge and plastic traditions, is being torn asunder by global capital. The existence of local communities untouched by modernity and alienated from the illegitimate national community is perceived as offering a critique to the language of universality, Western modernity, progress, science, and development (Nandy, 1983).

Gandhi is supposed to have offered the most powerful critique of the instrumental conception of Western society and particularly of the nation-state.[24] In its place, he put forth the concept of society based on Dharma. The arrival of the colonial state facilitated the inauguration of the national imaginary.[25] Gandhi's anticolonial consciousness was civilizational in its scope and breadth. Even the earlier nationalists had already laid out the nonsecular basis of this community.[26] Gandhi developed this idea further and reworked the moral basis of this constitutive community. In his imagination, performance of duty took precedence over the exercise of individual rights.

It was against the grain of such an imagined community that the resentment of the nascent Dalit formation was voiced in the early twentieth century in the form of demand for civic rights. In the midst of pervasive poverty and social exclusion, the entry of Harijans[27] into temples, schools, and other public places absorbed the attention of nationalist Congress leaders. The famous Gandhi-Ambedkar dialogue has to be situated within this historical context (Parekh, 1999). Gandhi used both moral and political arguments to engage with the issue of untouchability (Parekh, 1999). He argued for an internal critique of untouchability and pleaded the issue to be resolved from within the Hindu fold. For Gandhi, Hinduism has moral resources to absorb the stain of untouchability. The Indian leader launched a twofold attack against untouchability. While social inequality demanded activism to alter the real-life conditions of Harijans, moral inequality needed a critical reinterpretation of the *shastras*. For Ambedkar, Gandhi was putting forward an ill-founded thesis by suggesting that there is no internal connection between the caste system and the practice of untouchability (Parekh, 1999). Ambedkar resolved this by stating that though he was born a Hindu he would not die as one.

In addition to appealing to the moral conscience of upper-caste Hindus, Gandhi threatened them with the political consequences of a deep cleavage arising from the demand for independence for the nascent nation (Parekh, 1999). However, Gandhi was keen to keep the issue of untouchables outside the ambit of the colonial state. His interest in maintaining a distinction between social and political reform arose from the above logic. Ambedkar found this logic spurious hence his outburst against Gandhi's fast in 1932, which he termed a "filthy and a foul act" (Parekh, 1999, p. 229). Gandhi intuitively understood that the demands raised by the Dalit community were not focused merely on civic rights. He realized that they questioned the moral basis of this imagined national community. He appears to have tactfully reworked the question and discursively made the dispossessed the true bearers of this civilization; hence the attribution to them of the opulent term Harijans (god's own people). Against this background, Ambedkar moots the concealed character of the ascriptive society on which the nation is premised and suggests a severing of ties with this national community by asking for separate electorates. What have remained are Gandhi's sentiments, which are echoed by political theorists such as Kothari (1993) and Nandy (1983). This is similar to what the communitarians have taken on in the West:

> They attack the liberal notion of self, and relatedly liberal ideas of justice and citizenship, principally for their purported hyper-individualism and its forms of political obligation. The meaning and possibility of justice or democracy turns, in this view, on the identification of individuals with their community and its values. In practice the reach and appeal of community—as a theatre of governance, identity, rule and performance—stretches far wider. (Watts, 2004, para. 3)

This view is captured in the work of Indian communitarians such as Ashis Nandy. Thus, in the colonial period of the early twentieth century, the three main actors who determined the course of the formation of the untouchable community were the colonial state; Gandhi, representing the most charitable sentiments of the mainstream Congress party; and Ambedkar, revealing the supposedly hypocritical stance of the Congress party, including Gandhi. The issues that these actors dealt with ranged from civic rights to civilizational discourses, political rights to critical traditionalism,[28] and community rights ranging from looking at religion as a foundational premise to perceiving it as a source of privileged access to both the material and the spiritual worlds. Whereas Ambedkar disallowed the invocation of scriptural authority for the governance of the emergent society and fought against the curtailment of rights, Gandhi deeply revered and succinctly argued for the performance of duties[29] as ordained by the principles of Santana Dharma.

As indicated earlier, the next phase of the journey is the objective Indian nation-state. Nehru is associated with liberal, secular imaginings of the nation. The preponderant theme of the Nehruvian state was development. Set against the pluralistic vision of Gandhi's idea of community, Nehru's notion of development was viewed as a terrain where differences of all kinds could be dissolved and an unmarked modern citizen could arise. The most significant feature of the development rhetoric was supposed to be its inclusive and secular character.

In the globalized twenty-first century, the rhetoric of development as instituted by the interventionist Indian state has changed. From its beginning, the Indian state gained legitimacy by deploying the rhetoric of development to target the nascent national citizenry. Now, in the changed circumstances, many third world states do not even find it necessary to pay customary obeisance to the idea of the citizen or national development. While plurality and the nation are theoretically incompatible concepts in a liberal-democratic paradigm, current neoliberal regimes in some ways reckon with the plurality of the nation. Since the 1980s, issues addressed by the new social movements, such as gender, caste, race, and ecology, have come to occupy the minds of the political establishment around the world. These issues appear to have problematized notions of universal progress, development, and world history.[30] Yet this communitarian discourse seems to gloss over the asymmetric basis of this diversity or to deliberately refuse to address this issue.

Theoretically, the liberal worldview has been critiqued from several vantage points, including an essentialist communitarian perspective. While the essentialist communitarian critique of the liberal nation-state is opposed to Enlightenment modernity, in the third world it is very often also a parallel effort to retrieve the nonmodern. In this critique of liberal modernity, the nonmodern is supposed to resonate with communitarian virtues and values in many ways. In this communitarian critique, globalization is seen as a further threat to nationally marginalized communities. In the popular discourse, the question often posed is whether the supposedly Nehruvian scantily clad social insurance state served the interests of the nascent citizenry better or whether the neoliberal regime inaugurated by his techno-managerial grandson Rajiv Gandhi did the job better by ensuring a safe passage for the emergent modern citizens into their new selves. It appears that Nehru recognized the material and inherent social inequity but insisted that they be kept on the waiting list. Nehru's epochal "temples of modern India" were far off and few in number as the multitudes seethed as they watched the privileged worship these new gods. While establishing elite liberal institutions such as the Indian Institutes of Technology, setting up massively funded science laboratories, and building huge dams and giant steel factories, Nehru procrastinated over issues related to land reform, a universal educational system, and access to minimum basic calories from food grains. The liberal Nehru proved inefficient in managing scarce national resources, and his image as a liberal secular self appeared contentious and ambiguous to some.[31] Many social scientists have found a fit between material status and caste ranking in early post-colonial India and now.[32]

Negotiating Citizenship and Community Rights: Toward a Dalit Critique

There has been an attempt to situate the Dalit critique against the liberal, national modernity discourse and the Gandhian communitarian discourse. According to this critique, the idea of citizenship within the framework of liberal national modernity seems to be tied to deeper historical and cultural affinities. In fact, the unmarked citizen seems to represent only the upper echelons of the changing traditional society. This chapter interrogates this essentialist communitarian critique, which portrays the Dalits as a local community. The Dalits constitute a substantial minority of the Indian population. It is widely accepted that as a group they could not emerge as national citizens or enjoy the entitlements of development. Their painful precolonial/ national histories of bondage and social ostracism have continued without respite. The essentialist communitarian critique seems to argue that since the liberal state has failed to provide Dalits with citizenship entitlements, and the neoliberal state, which is aligned with global market interests, is supposedly accelerating the alienation of these groups, it is important to image the Dalits as a local community. This local community is then seen to be imbued with rich local histories and knowledge. It is in this manner that the Dalit worldview presumably poses a counter to the imitative national and Western modernity. Many NGOs, which have become crucial players after the withdrawal of the interventionist state, tend to work with these ideas of community development.

While acknowledging the broad strand of the communitarian argument, one needs to be cautious of the implications of completely doing away with liberal national spaces, especially for any historically disadvantaged group and specifically for a pan-national political community such as the Dalits. It is for this reason that this chapter posits an anti-essentialist conception of the community with a culturally sensitive notion of citizenship. Such a charitable hermeneutic may provide ways of engaging more concretely with the security and dignity of historically oppressed people. This effort in some way recognizes a need for the idea of development in the neoliberal age to forge a link between community rights and citizenry rights. To put it more bluntly, the chapter conjectures a politically potent position by offering to combine both liberal and communitarian positions on the question of a just order for the globally and nationally marginalized community known as the Dalits.

The Indian Constitution (Basu, 1987) is a liberal document that recognizes the individual as a political unit. At the same time, it acknowledges the Dalits as a community by suggesting legal ways of restraining institutional oppression and providing positive affirmatory measures.[33] While the implementation of such measures has generally drawn flak from state agencies, citizenship entitlements are yet to come for the large swathe of the Dalit population. Less than two percent of the twenty five million Schedule Castes and Schedule Tribes have boarded the middle-class bandwagon. Electorally, more than one-fifth of the members of Parliament are Dalits, but their effectiveness is thwarted by the absence of a substantial educated middle class among them.

Hence, my submission is that marginal communities of pan-national character stand to make a better political bargain only if they consolidate themselves as a national citizenry. This national citizenry is not similar to the unmarked citizen of the national modernity kind because, as the analysis has so far suggested, that unmarked variety always represented only the dominant castes/classes. Nor is it the completely noninstrumental variety that the NGOs profess. The NGOs emphasize the encumberedness of the individual and the need for community rights to prevail over individual rights. The communitarian variety loses its impact because it does not offer secular salvation, and without this the chance of attending to spiritual alienation seems dim.

To conclude, the Dalit movement appears to offer simultaneously a critique of a certain interpretation of Gandhi woven into the alternative discourse of the NGOs and a certain rendering of the liberal Nehruvian position. In effect, it takes on a particular communitarian critique and a specific liberal position.

More specifically, the critique seeks to problematize the liberal, secular self of nationalist leaders such as Nehru. In sum, the effort is to suggest through the critique the not-so-liberal position of the liberal, secular self and the vertical nature of the communitarian position. To repeat, this chapter has explored the possibility of combining an anti-essentialist notion of community with a culturally sensitive notion of citizenship. Otherwise, on the ground, development with dignity would continue to evade Balaram, the Dalit driver. Somewhere, a civil society organization such as DDS has become a source for incivility toward its own employee.

Notes

1. The Constitution of India recognizes the former untouchables and *adivasis* as Scheduled Castes and Scheduled Tribes, respectively. The decennial census in India enumerates the Scheduled Castes and Scheduled

Tribes separately. Other groups are not enumerated on the basis of their caste.

2. The investigation was led by the well-known human rights activist K. Balagopal. The report categorically indicted the NGO for its undemocratic functioning and recommended the revamping of the board of directors, who were mainly drawn from the upper caste.
3. KNPS was established in Vijayawada in 1998. It is interesting to note that this organization was a result of the People's War Group (the Maoist wing of the Left Party in India, which believes in armed struggle against the state) conceding to the demands of the Dalit cadre within the organization. However, the resolutions of KNPS included struggles primarily against the growing market reforms (privatization) and the cuts in the state welfare budget, along with the punishment of culprits perpetrating atrocities against the Dalits.
4. The former prime minister V. P. Singh proposed proportional representation in government jobs and the education sector for backward classes in accordance with their population on the basis of a report by the Mandal Commission in the late 1970s. This was deeply resented by the forward castes, and there were many anti-reservation protests across the country, popularly known as anti-Mandal riots.
5. In his book *Of Grammatology*, Derrida (1976) demonstrates the nonderivativeness of writing. In this connection, he observes, "grammar is independent of semantics and phonology" (p. 57).
6. This interesting phrase was coined by Susie Tharu in a different context. She mentioned it at a conference on the "Philosophical Foundations of Social Movements" organized by the Department of Philosophy, Pondicherry University, on September 2, 2005.
7. Much against Foucault's exposition of the microphysics of power and the appropriation of Foucault by South Asian scholars prophesying everyday resistances instead of transformative politics, power seems to operate in a raw and brutal fashion in third world societies.
8. When the former chairman of the Constitution Review Committee, Justice Venkatachalliah, visited the University of Hyderabad to deliver a lecture at a human rights conference on January 13, 2000, various Dalit student organizations protested against BJP-led government designs to erase special provisions accorded to Scheduled Castes and Scheduled Tribes. Their slogan was "Bachao Bachao, Ambedkar Bachao, Hatao Hatao Venkatachalliah Hatao." (Save Us, Save Us, Oh! Ambedkar, Please Save Us! Remove, Remove! Let Venkatachalliah Be Removed!)
9. Professor M. Madhava Prasad, a well-known cinema critic, while commenting on identity movements, remarked that they symbolized an era of "expansion of reason," at a Culture Studies Conference at the Central Institute of English and Foreign Languages (CIEFL), Hyderabad, in 2005. Unlike Madhava Prasad, Aditya Nigam calls for abandoning rea-

son and locating the Dalits in a heterogeneous time rather than in the historical present of the linear world.

10. Dalit feminist theorists have questioned the patriarchal basis of this identity formation. Sharmila Rege (2000) has conducted extensive empirical and theoretical investigations to establish a Dalit feminist standpoint. This standpoint problematizes not only patriarchy but also the received feminist historiography in India.

11. See *Of Grammatology* by Jacques Derrida (1976). It is of immense significance that Derrida demolishes the distinction between natural and literal writing in the Western tradition. He states that rhetoric envelops both the literal and the figurative. In some sense, the Dalit cannot be constituted as merely the literal or only the figurative.

12. Notwithstanding the subaltern character of the campaign, it is still an elite representation of the event. People, who are representing, including the Dalits, cannot be grounded in the reality of the alleged victim. In that sense, the stutterers and mutterers also constitute an elite formation.

13. Two varieties of understanding inform scholarship in this area, one of the McLuhanian kind and the other of the Derridian kind. The McLuhanian kind is logocentric, very speech centred; while the other problematizes the logocentric approach. For McLuhan, writing is visual code for speech. Writing imitates speech. Derrida reverses the connection between speech and writing. This distinction has the potential to produce an internal historiography of media. For an extended discussion on McLuhan's idea, see his *Understanding Media*. Similarly, see Derrida's seminal text, *Of Grammatology*.

14. The Dalit audience for state radio and TV may be higher, but other parameters remain the same in these media also. There is some marginal representation as far as South Indian cinema is concerned in terms of actors and directors. In literature, autobiography appears to have become almost a Dalit genre. I find the monopoly of this genre very problematic. While the upper caste produce literature in a variety of genres and take on the moral responsibility of telling "our stories," the Dalits through the autobiographic mode are allowed to tell only "their stories."

15. Even when one argues that the Dalits possess a precolonial and a prenational memory, it is still possible that this memory is revitalized through capitalist modernity. Positing the Dalits as an administrative category need not preclude their historical and specific forms of reasoning. Their new-found homogeneity is an artifice necessary for building solidarity in times of massive fragmentation and differentiation.

16. The email sent out through the anonymous group Spanadana accused Sateesh of beating his Dalit driver with his footwear and later using his position to elicit an apology from the victim himself.

17. See http://www.ddsindia.com; accessed March 23, 2006.

18. The email from KNPS pointed out the closed nature of the NGO groups operating in the country. The anonymous email, which circulated only in NGO circles initially, was not available to others even after repeated demands from the Dalit activists.
19. Here is the full text of the letter that the Secretary of KNPS wrote to DDS:

> Sir,
>
> I am writing this letter as a Dalit activist. I know that you are a well-known public figure in the NGO circles. I hope you take my letter in the right spirit. I have been informed by some reliable friends that you had beaten up a Dalit employee of your organisation with your chappals. This came as a shock to me. Even the Feudal lords are scared to beat up Dalits these days. How did you do this? With what authority did you do it?
>
> When this issue came up in your staff meeting, it seems, you had agreed that you beat up the Dalit employee for telling you lies. You committed a crime punishable under SC/ST ATROCITIES (PREVENTION) ACT 1989. The concerned individual may not have complained to the Police for reasons well known to us. You are a responsible public person. What did you do about it? I want to know what punishment you had awarded to yourself for this crime.

20. This mail was addressed to KNPS and another Dalit human rights group known as Sakshi, which coordinated a fact-finding committee. The committee included a well-known human rights activist, K. Balagopal.
21. Letter from DDS to KNPS; private mail to KNPS state secretary, K. Satyanarayana.
22. The email piece was titled "Of Pada Puja and Local Communities."
23. This position offers a critique of both communitarian and national modernity. In fact, some scholars using this approach have questioned the subalternness of subaltern historiography.
24. Most scholars on Gandhi find his *Hind Swaraj* (1938) a seminal text in profoundly critiquing Western civilization. Anthony Parel (1997) has a long preface appended to his edited text in *Hind Swaraj and Other Writings*.
25. Even as scholars such as Sudipta Kaviraj (1993) and Partha Chatterjee (1992) demonstrated the ingenious efforts of the native intelligentsia in imagining a national community, they held the view that colonialism was crucial for its mapping.

26. Both essentialist scholars such as Ashis Nandy (1983) and nonessentialist theoreticians such as Partha Chatterjee (1992) have amply suggested the nonsecular basis of the intelligentsia's imagined nation.
27. Gandhi used the term Harijans to empathize with the untouchable community. This form of address has been criticized by leaders of the untouchable community.
28. Gandhi opposed the provision of separate electorates given to the untouchable community by the Mcdonald Award. Ambedkar criticized Gandhi for taking this position.
29. Bhiku Parekh (1999), elaborating on the variegated nature of the encounter with colonialism, remarks, "For the modernists, Hindu society was beyond hope and its salvation lay in radical reconstruction along modern or European lines. The critical modernists or syncritists pleaded for a creative synthesis of the two civilizations, whereas the critical traditionalists preferred to mobilize indigenous resources, borrowing from Europe whatever was likely to supplement and enrich them" (p. 42).
30. See Nandy (1983), In his classic *Intimate Enemy: Loss and Recovery of Self under Colonialism*, Nandy wrote in the aftermath of the Second World War and Vietnam that "it has become more and more apparent that genocides, ecodisasters and ethnocides are but the underside of corrupt sciences and psychopathic technologies wedded to new secular hierarchies, which have reduced major civilizations to the status of empty rituals" (pp. ix–x).
31. It appears that Nehru was not particularly happy about the positive affirmation policy and he was disturbed about Ambedkar's drafting of the Hindu Marriage Bill in 1955.
32. Summing up the conclusions of a cross-section of social scientists, G. Aloysius writes, "the lower castes in general are poor and the underdeveloped and the lowest castes are the poorest" (Unpublished). Jean Drèze, in a 2005 lecture at the University of Hyderabad, while speaking on his work on Himachal Pradesh, revealed a connection between cultural formation and material formation, the visibility of Dalits, and the relation of this Dalit visibility to the overall development of the region.
33. Positive affirmatory measures in the Indian Constitution constitute the right to appeal against any form of untouchability, and articles pertaining to reservation in education and jobs in the government sector.

CHAPTER SEVEN

Spatial Narratives of the Local: Bringing the Basti *Center Stage*

Vinita Agarwal and Patrice M. Buzzanell

> It was five-thirty in the evening. I climbed onto the roof. My younger brother Rehan was flying a kite. Our roof is the lowest in our colony, and everyone else's is quite high. And some even have two- or three-storied houses. I spread a sheet on the roof and sat down. Our neighboring house is two-storied. On its roof, a boy was doing his work from school. He is Raju and studies in eighth class. I didn't use to speak much with him. And he also used to stay aloof. Because in my house, they are very strict about such things. Sometimes, when it couldn't be helped, or when we had some work with one another, we would speak, but he always addressed me as his elder sister.
>
> My brother's kite was flying in the direction of his roof. A wind was blowing. Many people, boys, children, were on their roofs. The sound of decks playing in different places could be heard. Also could be heard the sound of girls chatting and laughing. (Tabassum, A., Ali, S., Rai, S., Neelofer, Ayesha, Shahjehan, Bobby, et al., 2002, p. 179)

THE *CYBERMOHALLA* DIARIES capture the ordinary events and relationships among people living in transition in a temporary *basti*, or neighborhood, on the fringes of Delhi, India. *Basti* residents are migrants from many parts of the country looking for work or family connections. Many *bastis* eventually become permanent, as urban plans are redrawn to accommodate the influx of people. Amenities such as electricity and water and infrastructure such as roads, schools, and hospitals have slowly been built over the years.

While the *basti* space itself is in transition, so, too, are the writers of the diaries. These writers are mostly young girls at intersections of childhood and early adulthood, of intimate self-reflection and candid observation, of marginality and representation, of locality and displacement, of enduring and fleeting social relationships, of hope and fear. They are school dropouts or irregulars (in their attendance). Many do not attend school because of family responsibilities, marriage, violence in their locales, or other reasons. Drawing upon these young women's descriptions of their *basti* and the structure of their daily lives within it, our goal is to reveal the interrelationships between their spatial representations and their feelings and thoughts of displacement, transience, and violence.

To interrogate the communicative representation of spaces within the diaries, we draw from three different dimensions of interdisciplinary scholarship: from the postmodernist geographical construction of space-time, feminist architectural critiques, and cultural analyses of urbanization in the global age. We explore their integrative potential in revealing the gendered construction of the architectural forms of the diaries, reflecting on transient sites in a large urban city in India.

Geographical and Spatial Narratives in the Global World

As a critical look at the spaces constructed and revealed in the journal entries, our chapter first engages a postmodernist geographical stance to "reject the imposed binary, [to] deconstruct and disorder it, and force it open to a multiplicity of alternative choices" (Soja, 1996a, p. 1421). In creating empowering forms of transmigratory spaces, the notion of Othering described by Soja (1996b) emphasizes the process of disruptions and modes of experimentation with unexplored forms of representation. Such an exercise is essential to force open a window on the silent musings of the journals, which question, but only hesitatingly, and are at peace with their neighborhood. The specific ways in which these geographical spaces are constructed is critiqued in the work of the spatial feminists. Spain (1993), for example, looks at differential and gender-specific ways of negotiating public spaces, such as schools and workplaces, to access socially produced knowledge. Markusen (1980) has examined the influence of the single-family dwelling on gender segregation, and Wilson's (1991) work has focused on urban planning as a significant determinant in reinforcing binaries such as nature/city and male/female. By including feminist spatial critiques (see also Hayden, 1997; Kaup, 1997; Wyly, 1999), we bring to our analysis an exploration of the young female diarists' gendered spatial characterizations by revealing how space intersects with social practices, roles, and the events that unfold in their writings. Related to, and extending this line of work, the feminist architectural perspective focuses on the role of architectural design as being of itself productive of specific gender identities (Coleman, Danze & Henderson, 1996). Furthermore, we explore the social construction of time and space in the imaginary geographies of the journals (Harvey, 1990) and the rootedness of the local in the global or the geography of centrality (Sassen, 1998). In this geography, the margins of the urban spaces are pushed ever further out, even as the global elites occupy the cultural and economic center stage of the urban centers in the globalized world. Such transnationality of the globalized urban centers constructs, in turn, the disadvantaged margins, which sink ever further into the discontinuities and contradictions that gave rise to them in the first place. Appadurai's (1996) writings further allow us to look into the unique forms of violence, displacement, and victimization that define such margins.

Our exploration, then, is a collective journey into the discursive representations of the windows and the lanes in the journals, lying at the edge of a technologically defined space (i.e., *cybermohalla*) and creating an imaginary geography of narratives. Our journalists reinscribe their world with the remnants of their uprootedness and transient migratory identity, while also depicting what is routine and authentic in their identity constructions. We look within these worlds for the architectures they define, the surprising aesthetics of violence and hope interwoven within the narratives, and the fragile construction of social practices within the spatial characteristics of their neighborhood dwellings. This chapter is at heart an interpretive, situated inquiry that asks how particular forms of displacement and marginalization define the local in our global world, as illuminated by cyberjournals of young girls and women in a *basti*.

Notes on Our Method and the Journey

According to Bagchi (2002), the *cybermohalla* project "takes on the meaning of the word mohalla, its sense of alleys and corners, its sense of relatedness and concreteness, as a means of talking about one's 'place' in the city, and in cyberspace" (p. 177). The online diaries are a part of a local computer center called Compughar set up by the NGO Ankur in collaboration with Sarai in May 2001 using free software and low-cost media equipment as part of their experiments with alternatives in education. The diaries are published online at Sarai's Web site (http://www.sarai.net) as part of their *Sarai Reader* publication.

From the three sets of data, consisting of forty-two journal entries, we interrogate those that convey the often (perceived) bleak thematic binaries of the diarists' existence through a postmodern sensibility (Blunt, 1994; Weedon, 1997, 1999). While it contextualizes the local through the deterritorialized social, the *mohalla* attempts to restore the locality of the individual voices, experiences, thoughts, desires, and feelings to issues of social representation, spatial empowerment, and resistance voiced by postmodernist and womanist feminists (hooks, 1984).

We come to this analysis from different feminist and cultural positions. One has familiarity with the urban and cultural context of the cyberjournal project because of her long-standing childhood and educational ties as a mass communication research school graduate with the city of Delhi. The other brings a long-standing concern with feminist advocacy and representation of girls' and women's experiences in academic work as well as in technology design and implementation.

Following Harvey's (1996) argument that "to say that something is socially constructed is not to say it is personally subjective" (p. 212), we examine social reproductions that embody spatial and temporal representations of the online self in the journals. In doing so, we tease out the gendered articulations of space-time principles in the structure of the social practices that both guide and maintain fragile social order in the *basti*. In choosing a textual analysis of the journals, we underscore a belief that

> [t]he occasions, spaces, and modes of representations are themselves forms of power rather than mere reflections of power residing in the material "facts of life," and the "big structures," through which the power of the class, capital, or the state are expressed. (Smith, 1992, p. 495)

Furthermore, we interrogate the journals' discursive space to tease open their borders and thereby suggest critical occasions for thirding. The process of thirding described by Soja (1996b) envisions a journey that explores the creative potential of lived spaces as a mode of resistance to the oppressive practices of marginalization, transmigration, and gender. To explore the potential of trialectics is to examine what is unexplored; to indicate what was not possible through conventional practices and social processes. We hope that a radical postmodernist critique will shatter some of these silences, if only to hold out another alternative, to speak out in another voice, to reject the need for totalizing and essentialist modes of representation so deeply embedded in the journals.

Reimagining Geographies: Frames and Passageways

In asking how particular spatial forms of displacement and marginalization define the local in the globalized world, we found three main spatial themes referring to windows and lanes that enable the diarists to navigate the varied discursive and material conditions of their lives in the *basti*. These discontinuities and contradictions are embedded in the aesthetics of spatiality as the young girls and women construct their social practices and gendered identities in (1) windows signifying porousness of frames, (2) passageways implying transience and legitimacy, and (3) public venues allowing the witnessing and construction of violence.

Windows: Porousness of Frames

> The window sees many things on the outside. For instance, plants and trees, fights, some donkeys, electricity cables and poles, birds in the sky. She sees lanes. She sees a park, roads, dogs, and cats. She sees people

> coming and going. She sees houses being constructed, people on roofs, kites being flown, pigeons flying. She sees water coming into the tap. She sees the tap run out of water. She sees rain. The moon and the stars. A street lamp being repaired. Men walking around in a drunken stupor. Children returning from school. Things displayed in shops. The ice cream vendor. The juice vendor. The shops selling biscuits.
>
> The window sees on the outside two women fighting over water. And one is pulling the other's hair. And the window also sees two children walking past, singing songs. They are singing this song: yeh dosti hum nahin todenge [roughly translates to "we will not break this friendship," a popular Hindi film song]. The window laughs out loud. Then she keeps singing this song. Then she sees, suddenly, four to five girls are going somewhere. (Tabassum, A., Ali, S., Rai, S., Neelofer, Ayesha, Shahjehan, Bobby, et al., 2002, p. 179)

The window is an observer. She is mute but is able to watch; the window reveals but does not judge. She is passive, but absorbs, reflects, and participates in people's feelings. She lets in all aspects of the *basti* and shares in the interior lives of its inhabitants. She is amused by certain things, such as the two young children singing a friendship song. She observes the tap, with and without water, a dominant image in many *cybermohalla* journal entries. She sees that the tap occasions many rituals and interactions. In this diary passage and in others, she sees that the tap is the site of opportunity, ritual, and gender in its life-giving force, its marked times of opening and closure, and its association with the work of women, who bring utensils to catch the flow and use it in their homes. She observes that the tap is also a regular space for violence, with two women fighting on this occasion and other *basti* residents yelling, pushing, and crowding on different occasions. With the kites soaring high, pigeons flying, juice and ice cream vendors selling their products, people working, and girls and boys going about their lives alongside the neighborhood drunks staggering in the lanes, the window sees all. There is vitality to the neighborhood. It is full of the life and humdrum routine and regularities. While technically the *basti* may be only temporary, its web of social reproduction and commerce has taken on a vibrancy all its own, full of the new life of ongoing production, commerce, and trade. The window has the status of a permanent observer, an ongoing participant, almost a member of the neighborhood. The window literally and figuratively frames life in the *basti* and, for us, provides an entry view into the lives of those at the margins of society, space, and globalization.

Passageways: Transience and Legitimacy

> The lanes of this colony are such that if strangers, or someone's relatives venture in, they won't be able to locate the address they are looking for. They either return, or finally do reach their destination after an extremely long-winding search.
>
> If a couple of houses get constructed facing each other, what gets formed between them is a lane. People don't need to make the lanes. Lanes get made on their own. In this way, there were many broad and long lanes in the colony. But as population increased, the lanes became narrower. There is one lane here that is very broad. That is the lane with the masjid [mosque]. The rest are narrow and small. These lanes are named after the well known people who live in them. For instance Shaukat, Kallu, Chavva, Liyaat, Dulha, etc.
>
> Some lanes are clean, and some are dirty. What people really want is for the lanes to become straight, so visitors to the colony don't lose their way. Otherwise, they can never make out which lane they have entered and which one they have reached! People want lanes to have their own names or numbers, and for houses to be numbered in some recognizable order as well. Like it has been done in other places, like Seelampur, Gokulpuri, Bhajanpura. This done, it will become easy for people to find the house they are looking for. It will also make it simpler for us to give others our addresses.
>
> Once it happened that when my father was working at night, and I was doing my school work, we saw a man peering into the lane. My father asked him what the matter was. The man said he wanted to go to Aalam's factory. My father started to explain to him the way. Completely befuddled, he said, Let it be. I can't understand a thing. Actually, I do know the way. But I can't understand where I have reached. I think I'll go back to the road, only then will I find my way. I must have entered the wrong lane because it's dark. Saying that, he started to head back towards the main road (*Galiyon se/ By lanes*, 2002, p. 2).

The *basti* is a temporary neighborhood in the city of Delhi and does not have a formally identified urban status. As a result, residents find both a desire to name the lanes or have "recognizable order" so that they can give others their addresses and a desire not to have such designation because, once formalized, a lane becomes almost impassible as houses crowd the space. The fact that some lanes are clean and some are dirty, and that the lanes are formed in an arbitrary, haphazard way, is worth noting, especially in the diary entry that almost pleads for the lanes to become straight so that people do not lose their way. The lane, symbolizing the *mohalla*'s temporary and almost anonymous status, recurs through this narrative in the diarist's perceived need to have the lanes numbered and ordered, or in the lanes' contrast with the road, a legitimate space and route, from where the traveler enters the *basti*, the unknown, irregular, illegitimate space in its transience. At the close of this passage, an interaction with someone from outside the *basti,* the traveler, indicates the individuals can find their way into the *mohalla* from the road, but even if they know their destination, they get lost in the lanes, which curve and form in mysterious ways. In a sense, people can locate their goal or place only if they return to sanctioned and legitimate routes (roads). The logic of the lanes alludes to the transient and discontinuous journey of residents as well as logic of the *mohalla*'s construction, drawing its identity from the random construction of houses in a section of the city.

Public Venues: Witness to and Construction of Violence

> Turkman Gate: Here accidents mostly happen on roads. What I am about to tell you happened eight years ago. There is a school near Turkman Gate, G.G.S. Sec. School, Bul Buli Khan, Asaf Ali, New Delhi. A terrible accident occurred here in which one rickshaw walla [or "driver"] and three to four school children were injured and one child died as well. On seeing the accident, one woman had a heart attack and she also died. People stopped sending their children to school. Fear had filled their hearts.
>
> Gandhi Market: Here there is a place if there weren't a red light, people would become used to seeing accidents. But one happened here as well. Many years ago, the daughter of our pradhan [roughly translates to "local council head"] Roshan met with an accident that shook everyone's heart. This accident happened with a bus. People say that on coming under the bus, she died, but pieces of her brain lay fluttering, jumping around. Even today people tremble when recounting the accident.
>
> Opposite here: This road asks for one sacrifice every year, people say. Ten years back, there was a woman who was pregnant. She was

> crossing the road with her daughter when a bus came and the girl was run over. She came under the tire of the bus. She pulled hard at her mother and she fainted, screaming. And why wouldn't she have; I saw this with my own eyes. When I saw the girl, she didn't have eyes and her head had become flattened against the road. I can still see that face clearly and then the whole incident gets replayed in my head. Then police arrested the bus driver and friends and relatives of the woman broke the bus.
>
> A year back, two boys were listening to music while driving their car. They too met with an accident with a car. They lived in Turkman Gate. This had caused quite a sensation.
>
> It was during Ramzaan, before Eid. A young man was going towards Delite cinema hall after his evening prayers when he got run over by a truck. Lots of blood flowed out of his body. My younger brother had gone to see it. He told me.
>
> Whenever there is an accident around here, silence descends upon the colony and for many days, it is all that is discussed. (Azra as presented in Sarai Reader 02, pp. 181–182)

From the outset, there is a poignant current linking violence and silence in this diary. It speaks of the young female writers' perceived helplessness and agency, of their positionality as mere observers in the random acts of violence that befall *basti* residents. The horrors of blood flowing, brain pieces fluttering, and screaming children crushed by the weight of a bus are met with contradictory thinking, feelings, and behaviors. Residents are both afraid ("Even today people tremble when recounting the accident"; "People stopped sending their children to school. Fear had filled their hearts") and resigned (residents harbor such tragedy in memory yet accept such sacrifices as the price they pay for change; residents hope that symbols of official recognition, such as red lights, can put an end to tragedy, but accidents still happen). Some residents enact behind-the-scenes (relatives and friends breaking the bus) as well as city-sanctioned (police arrests) resistance. Residents carry on in silence, even to the point of having their horror silenced through a heart attack and sequestering children from school, but they feel an overpowering need to discuss the accident for several days and years afterwards. In this talk, the everyday tragedies as well as a sense of identity and community for the residents of the *mohalla* are (re)created, even while the starkness of their migratory status as temporary residents in a temporary *basti* is emphasized.

The road is a reminder of an ever-present undercurrent of violence: gruesome accidents whose victims are mostly young children en route from school, or walking with their mothers. In its ceaseless, omnipresent, motion, it is almost like a cruel deity that "asks for one sacrifice every year." The residents of the neighborhood, although resigned to the inevitability of its presence ("whenever there is an accident around here," not *if* there is an accident), grieve in unspoken silence for its victims while the young children carry the memory of the accidents along with their growing-up experiences ("ten years back," "a year back," "many years ago," "eight years ago"). The road is a symbol of inevitable progress, legitimacy, stability, and site of buildings and gates with names, but the road also symbolizes an acceptance of unpredictable and uncontrollable—yet regularized—violence and sacrifice of innocent children in the hope of progress. It is an endemic part of the migration to an urban environment, symbolizing much of what the city promises, embodied in the myth of the violent sacrifices it asks for in return.

Discussion: Negotiating Margins and the Elusive Center

In examining the perceptions of diary writers that coalesce around material constructions of windows, passageways (lanes and roads), and public venues (gates and roads), we locate liminal spaces, identity constructions, gender enactments, and feelings. In doing so, we find that the gendered characteristics of the windows and lanes reveal the gendered nature of relationships characterizing social order in the *basti*. For example, the selective feminization of the windows highlights the appropriate social role characteristics for women (passive, accepting, observant), which the girls embody in their lives. Our journey into the world of *cybermohalla* reveals that the writers of the diaries are as bound by the structural characteristics of the windows and lanes they observe as their materiality constrains and enables them in their daily lives.

Postmodernist spatial feminist critiques, such as those of Rose (1993), would characterize the narratives in the journals as occupying the transitional space within the marginality of many contradictory social, cultural, and relational borderlands constituting residents' existence as a temporary *basti* in the city of Delhi. The spaces reflect the many layers of peripherality/centrality shifts that face the Basti residents daily, such as those of: violence/silence, fear/hope, anonymity/identity, chaos/order, marginalization/legitimacy. With reterritorialization processes, these binaries are constantly disrupted and reestablished through the ongoing practices of representation embodied in the geography of the neighborhood. Thus, the lanes take on the character of their inhabitants, going silent when there is a horrific accident that kills a child or becoming busy, noisy thoroughfares during times of commerce (Hayden, 1997). The discontinuities of the margins define the local of this urban *mohalla* in the images of fluttering, jumping pieces of brain and crushed bodies in public sites, in the different constructions of displacement reflected in the inability to negotiate passageways, in the anonymity of a home without a recognizable address, and in the passive windows that take on the character of young female *cybermohalla* writers who observe but cannot change everyday happenings.

The female windows watch and absorb the ongoing processes of relocalizing practices within these immigratory urban spaces (e.g., routines of traditional gender; the centrality of water and the public tap), which are not altogether without violence and rupture, and the social order, which, although fragile, often draws on the traditional resilience of its gendered dimensions during this transitional phase. We would like to acknowledge, however, that our exploration was constrained by our own relative distance as observers from the real context of the Lok Narayan Jai Prakash (LNJP) *basti* where the girls reside. To that extent, our comments reveal the themes underlying the online space of the diaries, as presented on Sarai's Web site. A future study that combines analyses of online narratives with participant-observation of the project and of the girls' material existence would greatly enrich our engagement with the narratives.

Nevertheless, the narratives interoperate in interesting ways with existing feminist scholarship on spatiality and extend the gendered dimensions of social construction of space and time in significant ways. As an instance of narratives of transience, they poignantly capture Harvey's (1996) theoretical contributions on the intersection of social relations and spatiality. Although the diaries reflect a specific manner of appropriation of the physical architecture of lived spaces, they extend our understanding of the processes through which bordered spaces are assimilated into everyday practices during reterritorialization. They offer an intense and intimate look into the often violent and disruptive conflicts that underlie such transience and migration during the process of globalization.

Our journey into *cybermohalla* extends theorizing on spatiality and marginality, especially in the context of urbanization in our global world. While the girls' coming-of-age experiences seem characterized by their multiple and contradictory status as observers, participants, and maybe even victims of the push and pull of the larger social forces of chaos and order, control and vulnerability, death and hope, these experiences are by no means unique to the specific geographical and temporal context which they describe. The duality of modernistic binaries may well characterize the forces defining marginalization and displacement in contexts of urbanization—indeed, to some extent, in all contexts of reterritorialization. In contemplating the many nuances of Othering that the diaries describe, we are struck both by the need to engage more strongly with the marginalized spaces of displacement and by the hope and resilience embodied in their engagement with the spaces of everyday life.

CHAPTER EIGHT

Caste on Indian Marriage dot-com: Presence and Absence

Archana Sharma

OUR NEW GLOBAL society, which is characterized by greater flows of people, goods, and ideas across the globe, is part of a historical continuum. What is distinct about the contemporary global condition is the reorganization of postcolonial spaces in radical ways that has been made possible by technologies such as aviation, electronic media, the Internet, and the telecommunication revolution. This has led to a rethinking of traditional notions such as "community," "home," "homeland," "ethnicity," and "nation." The rise of electronic networks has made the Indian homeland a more immediate and interactive possibility for diasporic Indians such as myself.

Globalization and the collapse of the British Empire have led to the presence of unprecedented numbers of Indians in the West. The children of those who immigrated to Canada and the United States in the 1960s and 1970s have come of age, and a substantial heterosexual proportion of them are looking for marriage partners. In this chapter I focus on identity production in the age of globalization, where e-technologies such as the Internet offer diasporics (and their children) new ways of reformulating questions of identity. Those of us who were raised in North America learned about our home cultures through stories, music, and the films of Bollywood. But now the Internet offers a new community through technology that has the ability, in a sense, to recall home instantaneously. The "imagined" homeland, with its imagined voices, images, and sounds, is speaking back through instant messaging, email, Internet telephoning, and other Web technologies. How does having the imagined homeland alive and responsive affect how we identify as Indian here, in our adopted homes? This new technology has enabled India to be recreated within its diaspora. It is no longer a story heard but, rather, a diasporic story lived. This living, practicing, and recreating of the Indian self with the help of new technology is the terrain of significance on which this research focuses.

I am particularly interested in how Indians in the United States and Canada practice identity through marriage dot-com Web sites. This chapter is part of a larger project on Indians, the Internet, marriage, culture, and identity (Sharma, 2006). This chapter grew out of my examination of two Indian marriage dot-com Web sites, A1 Indian Matrimonials and Shaadi.com. The design of these Web sites enables specific identity performance. I examine the multiple ways individuals can (and do) identify themselves—through linguistic, caste, regional, and cultural identifiers.[1] In this regard I argue that these Web sites enable a rethinking of what it means to be Indian, and I examine how a desire to belong to a community is part of this process. Furthermore, I am interested in how the Internet organizes, sustains, or transcends traditional hierarchies of race, caste, class, sexuality, and gender for Indians in the diaspora.

I also believe that the techno-savvy, highly educated group of North American Indians should be studied because they represent a privileged segment of the Indian population in the United States. They are the model minority of America. A large portion of the Indian population is among the socioeconomic elite in the United States.[2] Statistics on Indians' education, occupation, and earnings in the United States suggest that Indians, like some other Asians, do fall into the "model minority" profile. That is, they are highly educated and employed in professional jobs. Most Indians in the United States are between the ages of twenty and thirty-nine, which also suggests that most Indians are young adults who are earning professionals. They are the cultural representation of what Indianness means in this context. Their privilege gives them an existing "voice" in the "melting pot" of the United States. It is imperative that this group of privileged hybrids be questioned on their identity production in terms of disruptive or productive possibilities and retrogressive practices. In other words, I do not see the use of marriage Web sites as a necessarily progressive or liberatory community or identity project by diasporic Indians.

Indian Marriage dot-com

Indian marriage dot com Web sites number in the hundreds, with some Web sites claiming to carry the advertisements of hundreds of thousands of Indians (http://www.Shaadi.com).[3] Most of them try to create as many categories or options as possible that speak to Indians while addressing the concerns of the multiple cultural communities of the Indian diaspora. These cultural practices are as diverse as astrological chart matching and the antiquated (but very prevalent) preoccupation with the light complexion of prospective brides (more so than grooms). Dietary preferences can also be specified, since this is an important aspect of Indian culture at home and abroad.[4] Web sites even allow users to specify "family values" and thus allow you to label yourself as having or find someone who has "traditional," "moderate," or "liberal" family values. These examples support and address the diversity of the Indian diaspora and the Subcontinent. They also point to the difficulty involved in creating and analyzing such Web sites.

An in-depth analysis of these sites and the occurrence thereon of certain categories (caste, class, skin tone, education, location) illuminates the kinds of practices of community on the Internet enabled by marriage dot-com sites. The categories I have chosen to pay attention to have greater explanatory power vis-à-vis the research question: How are community and identity practiced through marriage dot-com sites? For example, by examining caste identification, I am able to interrogate the negotiations of exo/endogamous marriage practices by Indians. By examining Web site options for occupations, educational status, and locations, I am able to investigate which markers of social status Indians value when seeking a marriage partner. These Web sites also describe the kinds of users that are interpellated by these sites. Certain categories, such as skin color or caste, also address the ways in which these Web sites organize, sustain, or transcend traditional hierarchies of race, caste, and class for Indians in the diaspora. New notions of Indianness are enabled by such virtual parameters and practices.

The Web sites, A1 Indian Matrimonials and Shaadi.com, do not exhaust the possibilities of identification or community imagining; however, as texts created within a diasporic culture they say something about this culture. In addition to examining the structural parameters of these sites, I also examined the profiles of fifty North American Indian men and women from A1 Indian Matrimonials.[5] I chose one hundred online profiles from A1 Indian Matrimonials because the parameters of its search engines enabled the most generic and broadest gathering of Indianness, so I assumed that it would generate the most diverse range of findings. Shaadi.com, on the other hand, has a search engine that is too specific, since it searches only within the user-specified parameters of language, culture, or caste. To get an idea of the kinds of male and female profiles on A1 Indian Matrimonials, I marked minimal parameters. That is, I searched for male and female users between the ages of sixteen and eighty-five who were located in North America and who identified their "culture" as Indian. I left blank religion and profession, the only other categories for a quick search.[6] Even with my minimal search criteria, my search revealed that the average age of my online profiles was thirty years old for men (range twenty-two to fifty-six). Most profiles were of men in their late twenties and early thirties. The average age of online female profiles in this sample was twenty-nine (range twenty-one to fifty-three).[7] The majority of the men lived in the United States. Out of fifty, only two men listed "North America" and five listed Canada, Vancouver, Toronto, or Montreal. The majority of women posting profiles lived in the United States, but there was a larger percentage of Canadian Indian women in these fifty profiles than in the fifty profiles of men. Clicking on each user name provided further details such as gender, age, marital status, height, weight, religion, culture, language, education, occupation, profession, location, and number of children. The profile also included a short description of the user, their "Preferences & Requirements," and what details they would like to know about a prospective browser in their initial response.

The analysis of both Web sites highlights the accepted hierarchic practices by Indians and any subversion of them (e.g., users who choose not to identify caste, income, skin tone, etc.).[8] In this chapter, I present an analysis of caste identification. Questions of concerns are: Who identifies caste? What kinds of castes are identified? How does the design of the Web site articulate caste? Which castes are over- and underrepresented? What emerges is an overdetermination of Hinduism among all the religious affiliations possible on these sites, and the hegemonic caste identity is "upper or clean castes."

Caste and Class in Indian Society

The Portuguese were the first to use the word "caste" to describe the web of kin relations in sixteenth-century Hindu India (Dirks, 2001; Paz, 1995). In the sense in which it was originally used, it had connotations of lineage, and the Sanskrit word that best describes it is *jati* (birth, species). The factors that decided one's caste are lineage, profession, territory, diet, and language. Members of the same caste thus live in a certain area, engage in the same work, speak the same language, and have the same religious rituals.[9] They also marry within the caste (endogamy), but only to families not closely related (exogamy). Caste is a support for individuals in turbulent times and brings with it a sense of belonging. To marry outside the caste is to lose one's caste status and put oneself beyond the pale of one's community.

The caste system has been correctly described as "a system that in practice relegates millions of people to a lifetime of violence, servitude, segregation, and discrimination" (Human Rights Watch, 1999). Under the leadership of Babasaheb Ambedkar, in 1956 more than 1 million Hindus of the lowest caste (Shudras) renounced Hinduism and embraced Buddhism. They chose to be called the Dalits, which means "the oppressed," or "the broken people." Today the Dalits constitute more than one-sixth of India's population, some 160 million people. Some are Christians, but most are Buddhist. It is worth noting that, since the substantial majority of lower-caste people are also poor, caste stands in for economic class. Dalits continue to live in extreme poverty, discriminated against and not allowed to claim land that belongs to them. With the exception of a minority who have benefited from India's policy of quotas in education and government jobs, Dalits are relegated to the most menial of tasks, such as handling dead animals, sweeping streets, working leather, and removing human excrement. Millions of Dalits work in agriculture, paid less than a dollar a day, or for a few kilograms of rice. They are routinely beaten up and shot by the police and higher-caste groups, and often Indian law enforcement agencies turn a blind eye to this abuse. Dalits are forced to sell their children to upper-caste moneylenders as bonded labor to pay off debts. Caste, class, and gender become an interlocking system of oppression for the Dalit women. They are often forced into prostitution to serve village priests and upper-caste patrons. Their bodies become a tool for control of the Dalit community, and for "teaching" the Dalits their place in Indian society.

The concept of caste is further complicated by the Indian system of four classes, and the caste and class systems have been at the roots of the enduring, and repressive, Indian social structure for more than two thousand years. The four-fold class division among Hindus complicates the 3,000 or so castes in India. Every Hindu belongs to the Brahmin (priestly) class, the Kshatriya (warrior) class, the Vaishya (merchant) class, or the Shudra (worker, peasant, servant) class. One's class is determined irrevocably at the time of one's birth. Like caste, the four classes have religious sanction. Indeed, their origin is described in one of the later hymns of the Rig Veda, dating back about two thousand five hundred years:

> When [the gods] divided the Man, into how many parts did they divide him? What was his mouth, what were his arms, what were his thighs and his feet called? The brahman was his mouth, of his arms were made the warrior. His thighs became the vaishya, of his feet the shudra was born. (Thapar, 2002, p. 125)

It is easy to deduce from this hymn that the four classes are arranged in a hierarchy, with the Brahmins being at the top of the heap, and the Shudras consigned to the bottom. This is indeed true in both the religious and the social senses. The three upper classes consider the touch of a Shudra to be polluting and will go to great lengths to avoid it. When contact does happen, it necessitates elaborate religious purification. Like members of the same caste, the members of a class practice endogamy (marriage within the class) and exogamy (marriage to families not closely related). Despite the fact that "untouchability" was abolished under India's Constitution in 1950, the practice of "untouchability"—the imposition of social disabilities on people by reason of their birth in certain castes—remains very much a part of rural India. In rural northern India transgression of caste marital practices can have dire consequences for the couple, such as social ostracism and physical violence (Chowdhury, 1996). These caste taboos are targeted toward women from higher castes who marry lower-caste men, since the onus of family and caste honor is borne by women (Chowdhury, 1996). Honor is tied to control over female sexuality and reproductive labor.

The caste practices are not exclusive to Hinduism but have bled into other religions, including Sikhism, which prides itself on its egalitarian principles. Also, these practices are cause for concern not only in the Indian Subcontinent but also in North America, as evidenced by the 2000 case of a young Vancouver Sikh girl, Jaswinder Kaur (Jasse), who was allegedly murdered by her parents for running away with a rickshaw driver, Sukhvinder Singh (Mittoo), in India. He was poor man from a lower caste and her family believed that she had dishonored them by marrying him. (For a fuller description of the events, see http://sikhsentinel.com/sikhsentinel0209/datelinenbc.htm.)

Caste on Marriage dot-com Web Sites

Edward Said's (1993) injunction informs the examination of marriage dot com Web sites: "In reading a text, one must open it out both to what went into it and to what [is] excluded" (p. 67). An in-depth textual analysis of one hundred online postings by diasporic Indians in North America on A1 Indian Matrimonials revealed a glaring absence. Not one claimed to be from a lower caste. No one claimed to be from the Shudra class. Further, the design of each Web site also indicated an overrepresentation of "high castes," with Brahmins having the most "options" to choose from. In other words, you can identify with multiple Brahmin caste and subcaste affiliations, whereas the category of OBC (Other Backward Castes) is the only representations of "lower caste" identification. This is replicated on the Web site Shaadi.com since the only lower castes that Shaadi.com recognizes are also OBCs. This category is actually a late twentieth-century entry in the sociopolitical divisions of India, the result of a political attempt to consolidate the lower-caste vote. The hegemony of Brahmins is as much in evidence on Shaadi.com as it is in the imagined Indian nation, both in India and in North America. For example, one glaring disparity is that it is possible to specify twelve different kinds of Brahmins on this Web site, whereas the lower castes are represented under one rubric. As further testimonial to the hegemonic power of Brahmins, this Web site does not list Dalits (the oppressed), also a late twentieth-century term, one that includes both Shudras (the lowest class) and those whom Hinduism has marginalized to the point of excluding them completely from the class system, variously known in India as the "outcastes" or the "untouchables." It is at this junction—the presence of high castes and the absence of low castes—that the production of community and identity is precariously located.

This is most evident on Shaadi.com, which allows you to search specifically for a partner in your caste community.[10] It is possible to choose from ninety different castes, a list that far from exhausts the actual number of castes in India. This preoccupation with Hindu upper-casteism is replicated in the online postings on A1 Indian Matrimonials. Of the one hundred profiles I viewed on A1 Indian Matrimonials, about 30 percent of Hindu males and 14 percent of Hindu females indicated their caste. Interestingly, however, only high or "clean" castes were identified.[11] Notice how even Indians often confuse caste and class, as when one user identified her caste as "Kshatriya," and when another user identified his caste as "Brahmin."

On A1 Indian Matrimonials and Shaadi.com community divisions are overdetermined by Hinduism and Hindu custom. This means that the caste system that divides the Hindu community is used to organize *all* religious affiliations. Such is the pervasive nature of caste in India that Shaadi.com lists the different sects of Christianity ("Syrian Christian," "Jacobite," "Catholic," "Protestant," "Evangelical," and "Born Again Christian") and the different sects of Islam ("Shiite," "Sunni," "Dawoodi Bohra," and "Memon") as "castes." This is another moment of Hindu overdeterminism. It is perhaps historical irony that traditionally both Christianity and Islam are known for their opposition to caste divisions. As a matter of fact the early proselytizers of both these religions gained a foothold in India on the basis of this opposition, and many Christian converts were originally from the "lower" castes.

The proportion of Hindu online profiles I examined on A1 Indian Matrimonials is significantly lower (72 percent of the women and 66 percent of the men) than the proportion of Hindus in India (80 percent). The category religion seems to be loosely defined by A1 Indian Matrimonials. Under the heading "Hindu," for example, you can choose Brahmin, Hindu-Jat, Kayastha, Kshatriya, and Vaishya. This is a conflation of class and caste, since Brahmin, Kshatriya, and Vaishya are classes, with various castes within each class. This might simply point to an everyday Indian reality, namely the overarching strength of caste affiliations among Hindus. More complexly, this may be a symptom of the sociopolitical reorganization in India during the past two decades. In the early 1990s India saw the violent advent of caste-based politics at the state level, which culminated in the fall of the federal government. This vote consolidation along caste lines was overlapped and later overshadowed by the rise of fundamentalist Hindu politics that sought to bring together the vote of the upper classes (Brahmins, Kshatriyas, and Vaishayas).[12]

My focus here is the practices of caste discrimination by second- and third-generation North American Indians. The online profiles I have examined suggest that there is some caste consciousness among second- and third-generation Indians in North America. As mentioned above, only 30 percent of Hindu men and 14 percent of Hindu women indicated their caste. It is noteworthy that the majority of users did not indicate their caste. In addition, one can also speculate on the gendered practice of caste affiliation. Are Indian women less concerned with maintaining caste communities? Or is their identification or lack thereof determined by the marriage practice of hypergamy, the tradition whereby a woman takes on the caste of the man and his family upon getting married? Research shows that within the Hindu religion there is a preference for caste hypergamy for women (Raj, 2003). My larger project details caste discrimination by examining how the mechanisms and negotiations of caste are played out for marriage dot-com users whom I interviewed, particularly for women (Sharma, 2006). Part of this negotiation is noted in the subversive practice of respondents who leave caste off their "desired" list when searching for their spouse.

There seems to be a rejection of rigid casteism by diasporic Indians, who live in an ostensibly egalitarian West.[13] Perhaps the rhetoric of the American Dream or notions of meritocracy in North America provide a virulent answer to the restrictive caste constraints of India. This rejection of casteism points to an ambivalent relationship of the diasporic Indian with India. For many second-generation Indians in North America, a return to India would be a deeply unsettling experience. An illustration of this ambivalence is found in the work of Toronto photographer Jag Gundu. His Web site (http://www.jagphotography.com) displays a 2004 exhibit featuring India, entitled "India, Love to Hate, Hate to Love." This title captures the revulsion, ambivalence, deception (going along to please parents), love, and desire many second-generation Indians feel toward a casteist India and its ramifications and residual effects in their lives. Gundu's words "my lens refused to deny its encounters" capture the essence of the encounter of the professional Indian with the reality of India. Second- and third-generation Indians who are of marrying age must negotiate their community production in the face of casteism.

The Risk and Promise of the Internet: Hindu EBBs and Dalit Web sites

It is difficult to ignore that cyber-communities emerge out of a process of self-selection that is governed by issues of connectivity, class, nationality, education, gender, and ethnicity (Morley, 2000). A key feature of the Internet is the homogeneity of the people who use it, which sociologist Julian Stallabrass suggests is the outcome of a process of "social self-selecting." This means that users of the Net are on the Net because they belong to a certain privileged ethnicity, class, gender, and nationality (Morley, 2000). The sad irony is that Internet users are lulled into accepting the "reality" of cyber-communities that have had gender, race, and class leached out of them, a construction that seems to invert the commonsense idea that it is people who make communities, not communication technology (Lockard, 2000; Silver, 2000). I make note of this process of self-selection because I understand the use of the Internet to find a mate for diasporic Indians as also a way to mark a self-selected community. On the one hand, this kind of self-selection flies in the face of the utopian rhetoric of cyberspace as a space devoid of race, class, gender, and, here, caste. On marriage dot-com sites, criteria such as class (profession and education), skin color, and caste are in fact used to find a spouse by many Indians. On the other hand, Indians on these Web sites are also narrowing their communities through self-selection. Therefore, I understand the Internet to be a contradictory place with possibilities. That is, the use of this technology by elite Indians living in North America has retrogressive and progressive possibilities. "The internet is complicated, at times contradictory in its potential and implementations, but still replete with possibility" (Kolko, 2003, 7). Saco (2002) calls this the "productive ambiguity of cyberspace" (p. 202).

To construct meaning from my Internet data, I relate them to other texts in two distinct ways. First, I have extended the traditional use of intertextual analysis by introducing an analysis of *other* Web sites. One set of Web sites that have been previously researched are the Hindu EBBs (electronic bulletin boards) that speak to a particular diasporic construction of the Indian nation as

Hindu/upper caste/patriarchal (Lal, 1999; Mitra, 1997; Rai, 1995; Sudha, 1993). Analysis of these Web sites also enables me to look at the intersection of nation and gender. I then juxtapose these Web sites with the Dalit Web sites, which give the reader a different construction of the Indian nation, one that deeply fissures any unified hegemonic Indian nation. I also deploy these Dalit Web sites to make sense of absences on A1 Indian Matrimonials and Shaadi.com along lines of caste/class.

In writing about the Internet and the possibilities of cyberspace, Ong (2003) examines the "the promise and risk" of the Internet. One of the risks of the Internet is the construction of a unified global ethnicity. Ong points out that "information technologies play a big role in engendering and channelling desires for a grand unifying project of global ethnicity that flies in the face of the diversity of peoples and experiences" (p. 88). She suggests that the material and social conditions within the local are more fruitful to examine than cyber-discourse that seeks to create a racialized diasporic identity—her warning is brought home when one examines the EBBs and the use of them by Hindu nationalists both in India and in North America.

The majority of users of burgeoning Indian religio-political EBBs are American diasporic Indians in highly paid jobs in North America (Sudha, 1993). The discussions on such EBBs promote a Hindu nationalism that demonizes its minorities (mainly Muslim citizens) and is intent on rewriting Indian history in order to reawaken a more masculine Hindu identity (Lal, 1999; Mitra, 1997; Rai, 1995; Sudha, 1993; Thompson, 2002). Alarmingly, Hindu militancy in India is funded by the (mostly American) wealthy Indian diaspora. Thompson understands this participation by the Indian diaspora somewhat differently. He suggests that this is one way for the diaspora to express "ethnic equality to other ethnic or majority groups" (pp. 414–415). This would put a softer face on these EBBs, which could be seen as the attempt of a minority group to express their identity in *opposition* to the majority white group that it is surrounded by. This kind of representation then seeks to put up a united front, if you will. What I am objecting to is that it suggests a unified Indian diaspora that mimes a national Indian identity along lines of caste, class, religion, and gender. It is part of my agenda to problematize this hegemonic construction of diasporic India.

There is opposition to the bolstering of such a unified global ethnicity (Lal 1999; Thompson, 2002). This was brought home to me when I examined the use of the Internet by Dalits. When I searched the Web using "Dalit" as the keyword I came across quite a few Web sites. It became apparent that the subaltern was speaking, or being spoken for, in a variety of ways, making connections across the globe using the Internet. One of these Web sites was http://www.gfa.org (Gospel For Asia), a Texas-based organization that proselytizes, fund-raises, and endeavors to raise awareness about the plight of Dalits. On this Web site two vulnerable groups of Indians (Christians and Dalits) have come together under one umbrella. Another Web site, http://www.hrw.org (Human Rights Watch), emphasized how the Indian state protects perpetrators of violence against Dalits. The Web site identifies Dalit women as the most vulnerable group. It also provides a link to the United Nations High Commission on Human Rights for reporting acts of discrimination, wrongful arrest, violence, and extra-judicial executions. Another Web site, http://www.idsn.org (International Dalit Solidarity Network), had a unique approach in which they interpreted "caste" to include people around the world who suffer discrimination on the basis of occupation.[14]

From the point of view of disrupting the idea of a hegemonic Indian nation, the most remarkable Dalit Web site I came across was Dalitstan.org. The name of the Web site invokes the concept of "nation" as a unifying political entity (as in "Pakistan," "Afghanistan"). It also simultaneously speaks to the desire for secession by an oppressed group, gesturing to an imagined nation (as in "Khalistan," the name chosen by separatist Sikhs in the 1980s). The Web site designers also complicate the definition of "ethnic" in their description of Dalits as "one of the most oppressed ethnic groups in the World" (Daslitstan.org home page, author unknown) perhaps in an attempt at solidarity with the fights around the world based on ethnic discrimination. This reading is bolstered when one looks at the comparisons the Web site draws with

Jewish struggle when it refers to violence against Dalits as "genocide" and references "the Sudra holocaust." A further connection to racism is made explicitly when the Web site refers to Dalits as the "Black Untouchables of India." Most relevant to my own study is the Web site's implication of diasporic Hindus in North America in funding the forces of fundamentalism in India ("Hindu Terrorists Made in America," link on Dalitstan.org home page).

I read these Dalit Web sites contrapuntally (together, and in contrast) with the Hindu EBBs. What emerges is a challenge to the notion of the diasporic Hindu in the West as a "modern" and "progressive" subject. These are voices of dissent and fissure within Hinduism, challenging the hegemonic Hindu subjectivity. My project is to mobilize Brah's (1996) notion of the Indian diaspora as a contested, heterogeneous entity, one that is fissured along lines of race, class, caste, nation, and gender.

Discussion

Poster (2001) examines how the production of self is possible through the Internet, and how the Internet is different from other media in this regard. He points out that, unlike print media or television, Internet communication forces subjects to position themselves linguistically and self-consciously. In the very act of Internet communication the subject is inventing himself or herself, or, as Poster puts it, "internet discourse constitutes the subject as the subject fashions him or herself" (p. 266). This notion of identity production is consistent with Hall's (1990) notion that identity is always becoming as well as being. This self-construction is significant because I interpret the process of Indians using the Net as one of identity production. However, in this instance, they do so within the confines of Web site designs and Internet communication.

Poster's (2001) notion of self-conscious positioning speaks to online profile postings as self-conscious positionings that linguistically gesture toward what is understood as Indian, American, or Canadian, hybrid, cosmopolitan, etc. Therefore, how caste is present or absent is also a part of this kind of identity production. Now, how one identifies is of relevance if we understand such self-positionings as part of a larger debate in the diaspora and on the Net. Thompson (2002) examines research on home Web pages that suggests that "new communication technologies are used to express and make public individuals' sense of self," noting that this kind of use of the Internet is a "far cry from sociological notions of a national mass media that reaches out to create a common national culture and sense of identity with the nation-state" (p. 411). Therefore the Internet is being used to create multiple voices and multiple publics who have an interactive part in producing themselves in this space while also using the technology for their specific communal, personal, national, or diasporic concerns.

Information technology and marriage dot com sites become enablers of a public discourse about issues of identity and community production. Online postings, online caste categories and identification, as well as those users who choose not to identify with caste affiliation begin to create a public discourse about how Indians in North America are negotiating, negating, and perpetuating the repressive legacy of casteism. Habermas (1962) writes, "the public that read and debated this sort of thing read and debated about itself" (p. 43). If the Internet, unlike other mass media, is able to constitute a public, in the way Habermas intended, then I argue that marriage dot-com Web sites are the loci where the Indian diaspora has an opportunity to debate about itself. This is especially true since, for Indians, seeking out an Indian partner *means* seeking out community. For many Indians, getting married is not about two individuals but rather about two families, or communities. Marriage is as much a commitment to a life partner as it is a practice to maintain cultural, caste boundaries, and to reproduce these communities (Shukla, 2003).

Saco (2002) reminds us that cyberspace may free us from the dictated movements of bodies, but it also "follows certain orders" and "it is the nature and impact of this different ordering that must be researched more carefully" (p. 202). It is this conceptualization of "ordering" that reveals the significance of marriage dot com Web sites that enable identity and community production. That is, these Web sites offer the possibility, through the practices of users, of a different kind of ordering of community and cultural practices. And yet, they may also produce older ordering of communities such as casteism, East versus West, and the North/South divide. For example, when caste and subcaste identification is possible on a Web site, one might argue that traditional hegemonic practices are being reinscribed. However, keeping in mind the dynamic interface between the user and hypertext, I am aware of the dual nature of discourses of resistance and compliance. That is, the Web sites might limit choice (by promoting some over others) in their structural parameters, but how people choose to engage with the technology is a process of negotiation worth investigating. I do envision the Indian users of marriage Web sites as Bakhtin's "intentional hybrids [who] create an ironic double consciousness" and are "internally dialogical, fusing the unfusable" (Werbner, 1997, p.5). Although, intentional hybridity may capture the practice of ethnicity/identity, we also need to interrogate the roots of discriminatory practices by diasporic Indians before ethnicity can become "a legitimate force of resistance" as Stuart Hall (1990, p.13) suggests.

In conclusion, I have shown how matrimonial Web sites allow a fusion of tradition with technology in which there is a simultaneous opening up of possibilities within highly specific cultural constraints. The Indian diaspora in North America is in the process of reformulating the concepts of "Indianness," "homeland," and "community," and I contend that the directions of this reworking could bear close scrutiny by anyone who is interested in questions of equity, social justice, and hegemony. My findings indicate that caste practices are being replicated in the design of Indian matrimonial Web sites. However, the nature of the Internet allows the user to accept or reject the practice of casteism. To what extent diasporic Indians actually replicate or reject the caste system in their real-life marriage practices is a question that will need to be addressed in future studies.

Notes

1. Arguably, these diasporic respondents on the Internet are a privileged group, but they are also dispersed peoples whose lives are complicated by their location, their race, their class, caste, and gender.

2. Asian Indians are among the most educated and professionally employed populations in the United States. Sixty-four percent of Asian Indians had at least a bachelor's degree, compared with 24 percent of the total population. Seventy-nine percent of Asian Indian men participate in the labor force, compared with the national figure of 71 percent. Almost 60 percent of Asian Indians were in white-collar occupations. The U.S. Census 2000 labels these jobs as "management, professionals and related occupations." Compared with all other ethnic groups, "Asian Indian men had the highest year-round, full-time median earnings ($51,900)" (Reeves & Bennet, p. 15). Along with Japanese and Chinese women, Asian Indian women had the highest median earnings among all ethnic groups, including whites. "They also earned between $4,300 and $8,800 more annually than all women" (Ibid.).
3. In fact Shaadi.com now boasts more than 1 million members. This increase (from 500,000) has occurred in less than a two-year time span. Such a proliferation furthers the importance of research into this burgeoning Internet phenomenon.
4. Orthodox Hindus do not eat beef, Muslims do not eat pork, and most Brahmins are vegetarians.
5. The male and female profiles were downloaded on January 17, 2003.
6. If you just want to "peek inside" as a casual browser you are allowed to search only within limited categories. The questionnaire for an advanced search is much more detailed.
7. Averages for men and women are calculated from the one hundred profiles I selected.
8. There are also marriage dot-com companies who provide space for alternative identities such as gay and lesbian Indians. In my larger project and here I have focused only on Web sites catering to heterosexuals.
9. This is complicated somewhat by urban Indian living conditions.
10. Some of the "castes" listed are truly mystifying, including, as they do, a region ("Punjab"), two religions ("Jainism" and "Sikhism"), and even a country ("Nepal").
11. The nine "castes" identified within the profiles were seven Brahmins, one Kshatriya, and one Hindu-Jat.
12. These constructions of caste leak into diasporic consciousness. Diasporic Indians have allowed caste to seep into their own communities, even if only to reject it.
13. E.g., the Burakumin of Japan, the sab groups of Somalia, the occupational caste people in West Africa, and the Paekjong of Korea.
14. The militant arm of the Dalit movement called themselves "Dalit Panthers," inspired by the "Black Panthers" of the United States.

CHAPTER NINE

The Virtual State of the Nation: Online Hindu Nationalism in Global Capitalist Modernity

Rohit Chopra

IN AN ESSAY ON GLOBALIZATION, Giddens (2003), commenting on the use of "new communication technologies by fundamentalist groups," observes that "Hindutwa militants have made extensive use of the Internet and electronic mail to 'create a feeling of Hindu identity'" (p. 50). The ideology of Hindutva, or Hindu nationalism, first formulated by Savarkar in the 1920s, proposes that India is essentially a Hindu nation defined by a Hindu cultural ethos that has existed for millennia and that non-Hindus, especially Muslims and Christians, are outsiders in the Hindu nation. Several scholars have noted that a strong bastion of support for Hindu nationalism is a diasporic, largely U.S.-based, Hindu population comprised in no mean measure of Indian technocratic professionals working in the global capitalist economy (Lal, 2001; Mazumdar, 2003; Rajagopal, 2000). The Internet appears to be the chosen medium for nonresident Indians (or NRIs) in general and those working in technology in particular to profess their affiliation with Hindu nationalist organizations, such as the Vishva Hindu Parishad (VHP, literally the World Hindu Council), and their adherence to the ideology of Hindutva. Das (2002) notes that

> a large section of the well-off, even "yuppie," NRIs have extended support to the VHP in a curious assertion of postnationalist ethnic identity. The

> principal mode of assertion of this support is the Internet, where...spurious "cyber-history" [has] been let loose in the service of "long-distance nationalism." (p. 44)

Mazumdar (2003) expresses a similar idea: "The much touted thousands of computer and software experts," who count among recent Indian migrants to the United States, "help maintain over five hundred VHP Web sites (www.vhp.org) with their messages of Hindutva, Hindu history, and Muslim-bashing" (p. 240). Rajagopal (2000) also observes that "with the proliferation of software engineers from India, the internet has become a site for expansion" of Hindu nationalist discourse (p. 484). According to Lal (2001), many Hindu nationalist Web sites are produced by politically and culturally conservative Indian software programmers in the United States. As the comments of these scholars suggest, Hindus sympathetic to the ideology of Hindutva appear to be using the Internet and cyberspace to reimagine the Indian nation in particular and specific ways.

At first glance, the content of Hindu nationalist sites—several of which are produced in a transnational diasporic space—could not more contradict the views of Jawaharlal Nehru, one of the architects of the postcolonial Indian state and its first prime minister. Indeed, Hindu nationalist discourse is well known for its antipathy to Nehru and his policies. In contrast to Nehru's understanding of India as a secular multireligion nation, Web sites such as Hindu Unity (http://www.hinduunity.org) and the Hindu Universe (http://www.hindunet.com) contain content describing Islam and Christianity as alien religions forced on Indian society through Islamic invaders and European colonizers. The emphasis on astrology and Vedic sciences in portals such as the Hindu Universe is radically at odds with Nehru's view of science as a quintessentially rationalist enterprise opposed to religion, ritual, and superstition. And the emphasis on the virtual consumption of products branded as Hindu as a means of affirming one's cultural identity—seen, for instance, in the Hindunet Signature Merchandise Web site (http://www.cafepress.com/hindunet2), which offers its viewers Hindu-themed products—might appear strangely unfamiliar to anyone who grew up in preliberalization India.

However else national identity might have been expressed in social practice and state action in India before economic liberalization in 1991, it was certainly not considered an object to be marketed along the lines of a business transaction. These characteristics notwithstanding, the specter of the Nehruvian vision of India is not absent from online Hindu nationalism, even as the core ideology expressed in pro-Hindutva sites is clearly drawn from the Hindu nationalist model of the Indian nation-state, articulated in Savarkar's book *Hindutva: Who Is a Hindu?*, originally written in 1923.

This chapter questions the premise that online Hindu nationalism stands for a total rejection of the Nehruvian model of the Indian nation-state. My argument is that online Hindu nationalism reflects at least two *general* characteristics of the Nehruvian model. First, in Nehru's thought as much as in that of Savarkar, the Indian *nation* is conceived of as a cultural ethos that is prior to and exceeds the Indian *state* that will come into being at independence. While Nehru defined the ethos of India as multireligious and pluralistic in opposition to the Hindu nationalist view of Indian culture as synonymous with narrowly defined Hindu culture, the general proposition that the nation exists as cultural surplus over the state is common to both models. This assumption permeates online Hindu nationalist discourse as a veritable article of faith, even as the cyberdiscourse defines Indianness along sectarian lines. Second, Hindu nationalist Web sites strongly stress the *scientificity* of Hinduism. It is true that Nehru understood scientificity as a secular quality to be implemented as part of a wider rationalist project in the state and in society and not as a religious quality. In addition, as the work of certain scholars (Nanda, 2003; Prakash, 1999) demonstrates, the current Hindu nationalist fixation with scientificity is rooted in nineteenth-century Hindu nationalist thought. However, Nehru also argued that India had once possessed the scientific temperament and that, relatedly, the quality of scientificity could be resuscitated in Indian society as the marrow of a national culture. In online Hindu nationalism, both these characteristics—the nation as cultural surplus over the state and scientificity as a national-cultural attribute—are refashioned into a narrative of the global Hindu-Indian nation that is optimized in cyberspace to function in a global capitalist economy compatible with the interests of an affluent diasporic Hindu population.

Toward substantiating the arguments above, this chapter proceeds as follows. It first details the models of the Indian nation-state proposed by Savarkar and Nehru, with a focus on the relationship between nation and state and the definition of scientificity in each model. The next section offers a close reading of two Hindu nationalist Web sites in the context of the preceding discussion to identify the modes in which elements of the Hindu nationalist Savarkariya and Nehruvian models are refashioned into an image of the global Hindu-Indian nation. The chapter concludes with some brief reflections on the implications of online Hindu nationalism for Hindu nationalist discourse at large.

Two Visions of the Modern Indian Nation-State

The views of Savarkar and Nehru about the nature of the modern Indian state may be read as two responses to the problem of a nationalism seeking to transition from the anticolonial to the postcolonial: how to endow the political institutions that were the legacy of foreign, specifically European, rule with a distinctive national, specifically Indian, character? In answering this question, both figures articulated a view on the relationship between the nation and the state. Both—although Nehru more than Savarkar—also addressed the role science could play in this transformation.

In the opinion of Savarkar (1969), the foremost duty of the independent Indian nation-state was to implement an overarching ideology as a principle of state: the idea of India as the Hindu nation. The "actual essentials of Hindutva" were "also the ideal essentials of nationality" (p. 137). The ties of "blood, birth and culture" that bound Hindus together as a people, and distinguished them from other peoples, would form the basis of a national state. As he asserted, in describing the Hindu claim to the Indian nation,

> a nation requires a foundation to stand upon and the essence of the life of a nation is the life of that portion of its citizens whose interests and history and aspirations are most closely bound up with the land and who thus provide the real foundation to the structure of their national state. (p. 139)

Savarkar, a revolutionary nationalist, shared Gandhi's objective of Indian independence, but strongly disagreed with Gandhi about the means to achieve freedom. In contrast to Gandhi's insistence on nonviolent struggle, Savarkar advocated violence as a means of anticolonial resistance. In opposition to Gandhi's insistence on Hindu-Muslim unity, for Savarkar, real freedom for India would be achieved not only by driving out the British but also by curing Indian society of the effects of Islamic invasions.

As scholars suggest, Savarkar's text *Hindutva*, written in 1923, in which he detailed his theory of the Hindu nation, may well have been written as a response to Gandhi's ideas (Neufeldt, 2003; Puri, 2001). Savarkar, as is commonly known, sought to define Hindutva in broader terms than just Hinduism as a religion. "Hindutva," Savarkar held (1969), "embraces all the departments of thought and activity of the whole Being of our Hindu race" (p. 4). At once a theory of origin and authenticity, Hindutva as the totality and essence of Hindu social existence stood for an ethos, culture, and worldview—a way of life—with its roots in the Aryan civilization of ancient India (Neufeldt, 2003, p. 136). This historical reality did not, however, extend to all the inhabitants of India: the conditions for qualifying as a member of this imagined community were, as Khilnani (1999) defines them, "geographical origin, racial connection (rather ambiguously specified), a shared culture based on Sanskritic languages, and 'common laws and rites'" (p. 161). Those who fit these characteristics belonged to the nation, while those who did not, such as Muslims and Christians, were outsiders (Khilnani, 1999, p. 161). But, if in theory Hindutva was proposed in universalistic terms beyond religion, in practice there was no way for outsiders to qualify as Indian except by rejecting their own faith.

Indeed, despite Savarkar's (1969, pp. 3–4) claim that the ideology of Hindutva was not primarily a religious doctrine, religion remained very much part of the equation. The test of loyalty to the Indian nation was the acceptance of India by Indian Muslims and Christians as their *holy land*. With their allegiance to alien faiths that had originated outside the cradle of Aryan civilization, Indian Muslims and Christians could never count as real Indians, that is, as culturally and nationally Hindu. However, even the acknowledgment of India as the holy land would not enable Muslims or Christians to qualify as cultural-national Hindus. As Neufeldt (2003) points out, for Savarkar, one could become Hindu

> through defense of Hindu culture, through intermarriage, or through adopting India as one's motherland or holy land. But there is one important qualification: To be Hindu one must subscribe to an indigenous dharma. A foreign dharma will simply not work. (p. 138)

The definition of a Hindu on the title page of *Hindutva* presents this conflation of religious, cultural, and national identity: "A HINDU means a person who regards this land of BHARATVARSHA, from the Indus to the Seas as his Father-Land as well as his Holy-Land that is the cradle land of his religion" (Savarkar, 1969). Vanaik (1997) defines Hindutva as a belief in a particular chain of reasoning, a discourse based on links between the four elements of ethos, religion, culture, and nation: "Ethos/spirit is at the heart of Hinduism, which is at the heart of, or coterminous with, Hindu culture (which defines Indian culture), which is at the heart of the Hindu nation (which defines the Indian nation)" (p. 151). Thus the question of religious identity, while subsumed under the category of culture in Savarkar's elucidation of Hindutva, functioned in effect as the main principle of division between Indians and non-Indians.

In Savarkar's view (1969), the ideology of Hindutva could be at once the inspiration, method, and rationale of the independent Indian nation-state. Despite the violence wreaked on their land by outsiders, especially by the Muslims and the British, the spirit of the Hindu people—as a cultural community—had endured, even if it had suffered severe body blows. Savarkar emphasized that independent India had to be a strong nation-state, one that would finally allow the Hindu ethos to flourish unhindered; the destiny of the Hindu nation-in-the-making that had been interrupted over millennia by successive invaders would be realized through the apparatus of modernity. Here Savarkar found a ready model in the writings of Guiseppe Mazzini, the Italian cultural nationalist (Hansen, 1999; Khilnani, 1999). As Hansen states,

> In Mazzini, Savarkar found an ideological framework and a political philosophy that combined cultural pride and national self-assertion with a modernist outlook and a vision of a strong culturally homogenous nation embodied by a unitary state—the vision of the making of a modern Italian nation-state. (p. 77)

Toward this end, Savarkar also strongly advocated compulsory military training for Indian students and the development of a military-industrial complex (Neufeldt, 2003, pp. 140, 142).

Establishing an independent Indian nation-state was not just a question of ownership of territory; the cultural unity of the Hindu nation was the real issue. Neufeldt (2003) points to the fact that "in principle Savarkar saw the principle of territorial unity as deeply flawed. It was, first, a product of the denationalization of India that had occurred through the introduction of Western education....Second, Muslims of India had never bought into the idea of a territorial unity" (p. 144). The modern nation-state, as Savarkar saw from Mazzini's writings, could be imagined in terms of cultural ownership. Territorial unity would be the spatial representation of this cultural unity, the empirical correlate of a self-evident historical truth that had survived as spirit. As Vanaik (1997) argues, the central feature of Hindutva is a "unity of culture and territory" (p. 150). In Savarkar's dream, the structure of the modern nation-state allowed for the successful realization of the Hindu historical spirit. The military-industrial apparatus would keep the flame of the Hindu spirit alive in the independent Indian state. It would secure and preserve cultural unity through the inculcation of qualities of aggression, cultural pride, and patriotism in the youth of the nation. Through the accumulation of force and techniques of self-defense and warfare, it would protect the Hindu nation from ever falling victim to cultural-religious invaders. India would undergo a profound transformation, yet, paradoxically, the transition to modernity would enable the truest manifestation of her enduring spirit from antiquity to the present.

For Savarkar, the power of the nation-state thus derived from its instrumental potentialities. The state was valuable for Savarkar precisely because of its ability to centralize force and its immense reach. As the hyper-powerful incarnation of the Hindu ethos, the Hindu nation-state would ensure the perpetuation of Hindutva. Science and technology, in the form of the military-industrial complex, would be the means for the Hindu state to perfectly realize its *telos*. This sentiment is expressed in Savarkar's twofold cry, his "message to Hindudom on the eve of this 59th birthday of mine...[was to] Hinduise All Politics & Militarise Hindudom!!" (Savarkar, 1967, p. 1).

Unlike Savarkar, Nehru (1995) believed that Indian identity could not be defined in narrow religious, cultural, or ethnic terms. However, Nehru too saw the state as a powerful agency with the ability to transform the lives of its inhabitants and science as a mechanism for social change. For Nehru, science *itself* would enable the transformation to postcolonial modernity, even as it would allow the new nation-state to manifest its unique ethos. Nehru's definition of science was not limited to scientific research and the apparatus for industrial development. Rather, what Nehru stressed was the quality of scientificity or the *scientific* method, which stood for a critical temper and a rational, enlightened outlook that could transform both the individual and the larger society. In *The Discovery of India*, Nehru observed, "The scientific temper and approach are, or should be, a way of life, a process of thinking, a method of acting and associating with our fellowmen" (1995, p. 512). As Prakash (1999) notes, Nehru

> was, as is well known, an avid advocate of what he called the scientific outlook....A much more expansive view of science than its conception as theoretical discoveries, Nehru's view projected science as a method that could be applied just as easily and effectively to order politics and economics as to understand nature. (p. 167)

Indeed, science was a torchlight that shone the way to freedom: "The scientific temper points out the way along which man should travel. It is the temper of a free man" (Nehru, 1995, p. 512). The scientific approach to social existence could, in this manner, explain and transform politics, society, and "life itself" (Prakash, 1999, p. 196).

It is undeniable that Nehru (1995) was strongly critical of what he saw as the deleterious effects of religion on society, especially in India. He did appreciate the need for spirituality in human existence, writing that "some faith seems necessary in things of the spirit which are beyond the scope of our physical world, some reliance on moral, spiritual and idealistic conceptions, or else we have no anchorage, no objectives or purpose in life" (p. 513). Religion, however, was the domain of "emotion and intuition" as opposed to rational thought (p. 513). Organized religion in particular was the very antithesis of science; it was guaranteed to lead to "narrowness and intolerance, credulity and superstition, emotionalism and irrationalism" (p. 513). Yet, for all its religiosity, Nehru saw Indian society and civilization as especially well suited to developing the scientific outlook, and, in fact, as placed at an advantage over the West in this regard. "The west," he wrote, "is still far from having developed the real temper of science. It has still to bring the spirit and the flesh into creative harmony" (p. 514). While India had "a greater distance to travel" to reach this goal, Nehru claimed that "there may be fewer major obstructions on our way, for the essential basis of Indian thought for ages past, though not its later manifestations, fits in with the scientific temper and approach" (p. 515). Nehru suggested that once science was introduced into Indian public life and consciousness, science and religion would harmonize with each other as compatible aspects of Indian culture. Thus, he confidently predicted, India would not be condemned to repeat the history of Europe: "The bitter conflict between science and religion which shook up Europe in the nineteenth century would have no reality in India" (p. 518), given that Indian society, in contrast to close-minded societies, was both historically and innately receptive to new ideas and change. It was such a radical and transformative scientific temper that, according to Nehru, needed to be adopted by the independent Indian state and cultivated in the people of India, toward realizing the promise of an Indian modernity (Brown, 2003, p. 189). In keeping with this emphasis, a state-driven national focus on scientific and technological development toward the goal of material progress had to be anchored in rational, appropriate philosophies of political and economic governance. Nehru's ideologies of choice in this regard were secularism and socialism.

Secularism, as opposed to a narrowly defined nationalism, appeared to Nehru to be the only viable political framework for India. Khilnani (1999), in his elegant evaluation of Nehru's secularist-nationalist vision for independent India, suggests that the genius of Nehru's conception of secularism lay in its radical inversion of any hegemonic idea of what it meant to be Indian. Nehruvian secularism proposed that "India was a society neither of liberal individuals nor of exclusive communities or nationalities but of interconnected differences" (pp. 171–172). This "anti-definition" of Indianness was based on Nehru's unshakeable conviction that "secularism was not a substitute civic religion, still less a political project to remoralize society by effacing religion and stamping a secular identity on all Indians" (pp. 177–178). Khilnani (1999) argues that this conception of secularism was not an imitation of Western models but distinct from them, a "model committed to protecting cultural and religious difference rather than imposing a uniform 'Indianness'" (p. 167). Indeed, Nehru's writings reveal recognition of the dangers of parochial nationalism found in the history of the West (Nehru, 1965, p. 166).

Rent by interreligious strife, India was also plagued by social inequality and poverty. The problems of inequity were, to a significant extent, derived from and maintained by caste hierarchies and feudal land relations, and could neither be seen as purely economic matters nor be solved as such. The economic imperative accordingly had to be yoked to initiatives of social justice. Nehru's answer to the problem was a humane socialism. Here as well, Nehru understood socialism as consonant with the ethos of Indian social life. Indian social life, in Nehru's view, had always emphasized the collective good over individual advancement; Indians would be historically conditioned and culturally socialized into psychologically internalizing the value of socialism and implementing it in practice. The Indian consciousness had not been contaminated with the ideologies of individualism and the profit motive to the same extent as in the West. "It would be absurd to say that the profit motive does not appeal to the average Indian," argued Nehru (1995), "but it is nevertheless true that there is no such admiration for it as there is in the West ... The Indian outlook, even of the masses has never approved of the spirit of acquisitiveness" (p. 522). He continues that "democratic collectivism," controlled by the state, would lead to "far more equitable sharing and a progressive tendency towards equalization" (p. 522). Collectivism, as Nehru saw it, was "fully in harmony with the old Indian social conceptions which were all based on the idea of the group" (p. 522).

Thus, in the Indian state imagined by Nehru, scientificity, as a worldview, would operate as the founding principle of the state. The independent Indian nation-state would stand as an incarnation of this principle. The state would promote an enlightened temper among its citizens, through commitment to the rational doctrines of secularism, progress, and development; the state-subsidized educational system would, likewise, stress scientific and technological education; the state would steamroller the Indian masses, urban and rural, into modernity through instruments and technologies of industrialization and development planning; and the centralized socialist state would accumulate economic capital and redistribute it on an egalitarian basis.

The similarities in the thoughts of Nehru and Savarkar are found in the logic according to which both conceptualized the relationship between the Indian nation and the Indian state. Both romanticized the exceptionalism of an Indian ethos, foundational to Indian society, on which the modernity of the Indian nation-state would have to be predicated. For both, the soil that nourished the nation-state would always exceed the nation-state as a cultural-historical surplus beyond its representations through the institutions of statehood. Nehru "contended that India's modern nationhood would be the fulfillment of its timeless unity as a community; India would be at once modern and not modern, at once represented and not represented in a state" (Prakash, 1999,

p. 225). For Savarkar, however, the cultural-historical base of India could be none other than Hindu. Equally, for both figures, science, as an instrument of material progress, would map perfectly on to Indian cultural traditions and would enable the latter to flourish within the framework of modernity. Of the two, Nehru, in fact, appears to be the stronger advocate of science. For Nehru, science stood for a *Weltanschauung* that could be grafted on to the collectivist and accommodating spirit of Indian society, expressed as much in the philosophies of socialism and secularism as in programs of industrial development. For Savarkar, science was primarily a concentration of force, an instrument in a project of cultural domination.

The Contradictory Legacies of Online Hindu Nationalism

Against this backdrop, I offer a reading of two Hindu nationalist Web resources, Hindutva and the Hindu Universe, along with related resources linked to each of these. The purpose is to highlight the two general characteristics of the Nehruvian and Savarkariya models of the nation-state—specifically, to explore how these ideas are expressed as elements of an imagined global Hindu-Indian nation. The Web resources differ in the mode in which Hindu nationalist sentiment is configured on each, yet there are strong commonalities and shared assumptions. Viewed together, they present an interesting network of similarities and differences that reveals the traces of the thought of both Nehru and Savarkar in online Hindu nationalism.

Searching the Web with the Google search engine for the term "Hindutva" displays the Hindutva Web site as the first link (http://www.hindutva.org). The home page of the site describes its focus as an analysis "of Current Global Politics and History from a Hindutva viewpoint." While the home page of the site contains a large volume of content on global events, the site also contains several other pages accessible through links in text or navigation. The four Web pages of the site, each representing a category of content, referred to here are the "Hindutva Pledge," "An Analysis of All Aspects of Hindutva," "Humanism and Rationalism," and "Rationalism: The Intellectual Bedrock of Secularism." Sudheer

Birodkar, one of the compilers of the Hindutva Pledge, is the author of the other pages.

The Hindutva Pledge merits special attention because of its attempt to define Hindutva as a rationalist, secular, and scientific doctrine. The pledge proposes to offer an authentic interpretation of Hindutva, representative of the "rationalist-humanist" outlook of Savarkar, in contrast to the dominant "spiritual-nationalist" take on Hindutva. The adherents of rational Hindutva

> pledge to dissociate Hindutva from all that is considered religious, casteist, sectarian, nationalistic and theistic; and pledge to retrieve the modernist, futuristic, egalitarian, secular, globalist, humanist and rationalist outlook which is the essence of the universal and timeless outlook that is inherent to Sanatan Parampara (literally 'timeless tradition').

The pledge also speaks of the principle of "the universal brotherhood of man." Also present here is a critique of religion, which is described as "a weakness of the human mind."

Astrology, palmistry, mythology, and the like are rejected as antithetical to true Hindutva. The pledge paints a utopian future where science is the ordering principle of society and where religion, largely relegated to the sphere of private practice, is kept firmly in check for the purpose of maintaining social order. In this imagined world, once the unruly variable of religion is "controlled" in public life, then can begin "the history of the human species, where science and quest for knowledge will dominate the human outlook and where WE as one species will face SPACE—the final frontier for humankind." Indeed, according to the Web masters of the site, the future of the human race lies not in "religion, spiritualism, dogmatism and religiously-inspired fanaticism"; rather, the "future Human being" of the brave new world structured according to the ethos of Hindutva "should be more attuned to frequenting Science Labs and Astronomical Observatories; rather than going to Churches, Temples, Monasteries, Mosques, Fire-altars, Synagogues, etc." The ideal human being of the future is a cyborg of a particular type, "a saintly scientist" who will defend the Earth both from terrestrial terrorism and extraterrestrial threat. The pledge recommends a universal code of conduct for all citizens of the future that is based on "qualities of reasoning and scientific research" and will help unravel the very mysteries of the universe.

The emphasis on a global project of inculcating the scientific outlook in human society aligns more with Nehruvian notions of scientificity than with Savarkar's instrumental use for science as a tool of national power. The critique of religion and the need for scientificity as a framework for governance and social relations likewise echo Nehru's views. The seemingly progressive sentiments espoused in the pledge are not dissimilar to the universal humanism espoused by Nehru. However, Savarkar's thought is by no means absent from the pledge; it jostles with Nehruvian reason in the mode in which an Indian cultural exceptionalism is invoked as the basis for the project of secular global humanism. The general claim here recalls the arguments of both figures about an exceptionalist Indian civilizational ethos, yet clearly tilts toward Savarkar's views regarding the substance of that ethos. Despite the disavowal of religiosity, Hinduism and Hindu culture are presented as the "universal and timeless outlook," on which the project of secular global humanism will be founded. While the pledge cleverly eschews using the term Hinduism as the inspiration for this outlook, describing it as *Sanatana Parampara* or the "ancient tradition of free-thought from India," *Sanatana Parampara* is but a euphemism for *Sanatana Dharma*, a term for the dominant tradition of Hinduism.

On pages of the Hindutva Web site, the Hindu cultural basis of the ideal world of the future is spelled out more categorically. Several Web pages— "An Analysis of All Aspects of Hindutva," "Rationalism: The Intellectual Bedrock of Secularism," and "Humanism and Rationalism"—begin with a section titled "What Is Hindutva?" In this section, in an argument redolent of Darwinian social logic, Hindu society is described as the only civilization "which has developed in its natural state, without interruption, since antiquity." It is "a prototype of *human* civilization" itself if that "civilization all across the globe had been allowed to develop in its natural state," the implication being that all other civilizations have evolved as aberrations against the natural grain of humanity. The "seeds of globalism and freedom of thought" can be found in two universalist "Hindu principles" articulated by the Hindu sages of antiquity "four thousand years before the world was to become the global village of today." These maxims are "This world is one family (*Vasudaiva Kutumbakam*)" and "The Universal Reality is the same, but different people can call it by different names (*Ekam Sat Viprah Bahuda Vadanti*)." Thus, the argument proves that Hindutva is "a timeless and universal compilation of human wisdom" and not a "narrow nationalistic doctrine."

Notably, the Hindutva Web site also presents issues pertaining to the Indian *state* as global concerns. The prescribed code of conduct for future citizens of the world spelled out on the pledge includes a "Common Civil Code for the entire citizenry of the globe" and a ban on evangelism. The reference to the common global civil code reflects the Hindu nationalist position on establishing a uniform civil code in India, according to which all religious communities would be governed by the same personal laws. The demand for a common civil code is periodically made by Hindu nationalists, primarily as a criticism of the Indian state's supposedly preferential treatment of Muslims and Christians in allowing these communities to follow their own laws in matters of marriage, inheritance, and custody. Also, the recommendation to ban evangelizing indicates the Hindu nationalist anxiety about Hindus in India converting to Christianity as a result of the efforts of powerful evangelists located in the West, especially in the United States.

Savarkar's thought also manifests itself in direct ways on the site. The Hindutva Web site quite explicitly denigrates Islam and Christianity and their followers. The "Motto" on the site's home page claims to promote "rational-humanism" over "violent monotheism," a response, it argues, especially warranted by the events of September 11, 2001. The real targets here are Islam and Christianity, in particular the former, and the fact is not even disguised. The home page is replete with numerous derogatory statements about Islam and Muslims, such as "terrorism resides in the hearts of all Muslims." The Web page on Islam in the Web site brims with similarly rabid sentiment. The Web page on Christianity concludes with the "logical inference that Indian Christians profess Un-

Indian beliefs." The secularism and universalism of the rationalist Hindutva project, therefore, do not extend to Muslims and Christians. It should be noted that the Web site is somewhat less virulent in its attack on Christianity, possibly because of its tactical support for the American "war on terror" and given that many Hindu nationalists are located in the United States. America is commonly perceived as a predominantly Christian nation in Hindu nationalist discourse. In effect, the discourse of globalism and universalism expressed on the Web sites enables the inversion of a parochial Hindu nationalism as the universal condition of humanity. The argument goes thus: If Hinduism stands variously for secularism, universalism, globalism, rationalism, and humanism, then those ideologies that are anti-Hindu (such as Islam and Christianity) are antisecular, anti-universal, antiglobal, irrational, and antihuman.

In my reading of the Hindu Universe portal below, I focus first on the configuration of globality as a Hindu-Indian technological, economic, and cultural practice on the Web resource, after which I identify the expressions of Savarkariya and Nehruvian thought. Searching the Web with the Google search engine for the term "Hinduism" displays the Hindu Universe Web site as the first link. A multipurpose portal, the Hindu Universe offers online visitors access to a range of social activities, cultural resources, and commercial products. The portal is a part of a cross-linked and sometimes confusing virtual global network of Hindu organizations. It is described as "the website for GHEN (Global Hindu Electronic Networks). GHEN is one of the many projects undertaken by HSC (Hindu Students Council)." On the same Web page, the Hindu Students Council is defined as a "voluntary run organization committed to realizing the ancient Vedanta truths such as *Vasudaiva Katumbakum (The Whole World Is One Family)*." The Web site of the HSC, in turn, describes the organization as an "international forum" dedicated to promoting knowledge of Hindu culture, although all the chapters of the organization are located in North America (http://www.hscnet.org). Offering opportunities for "self-development at the spiritual and professional level," HSC is "the first and only North American and international attempt by students and young professionals like yourself to explore, discover & experience the immense treasures of the time-tested knowledge and wisdom of the great Hindu culture." The text strongly resembles advertisements for new age techniques for acquiring management skills, where the truths of Vedanta are valued as much for their corporate value as their spiritual worth. If the world is a family, it is one of late capitalist modernity where membership derives from professional participation in the global marketplace.

There are two assertions here that are problematic as empirical and political claims. The first is the definition of North America as synonymous with "international," given that various populations of the approximately 20-million-strong Indian diaspora are to be found across the globe (Ganpati & Basu, 2004). The second is the fact that an organization based in North America, where the Indian-American population including Hindus and others, numbers 1.67 million (Dutt, n.d.) should arrogate to itself the right to speak on behalf of the faith, heritage, and religious practice of some 827 million Hindus in India, 23 million Hindus in Nepal, and anywhere between 15 million and 18 million Hindus elsewhere, assuming 75–90 percent of the 20 million diasporic Indians worldwide are Hindus (Office of the Registrar General, Government of India 2004; Central Intelligence Agency, Office of Public Affairs, 2005; Ganpati & Basu, 2004). In ignorance of sociological realities of the lives of Hindu communities across the world, the HSC site offers a prescriptive Brahmanical definition of Hinduism as a religion rooted in the authority of the Vedas. The claims of global reach and universal significance, then, are shallow but strategic, reflecting identification with North America as the center of the globe and with upper-caste Hinduism as the most authentic incarnation of the Hindu faith.

On the same portal, Hinduism and Hindu culture appear to acquire new meanings in the realm of commerce as well. The Hindu Universe portal includes a link to a Web site, Hindugifts (http://hindunet.tolshop.com), that is purely commercial and sells everything from gift vouchers to sarees and digital cameras to phones. The site may be considered the e-commerce section of the Hindu Universe portal, accessible independently or through the home page of the umbrella site. The Frequently Asked Questions page of Hindu gifts specifies that the site, which operates from Mumbai, India, is "an online store providing a worldwide service for sending gifts to India and for delivery of products in the United States." It also offers visitors the option to send items to other locations across the globe, and its products are priced in both Indian rupees and American dollars. Another site linked through the portal is the Hindunet Signature Merchandise Web site, which offers visitors a range of products with messages such as "Proud to Be a Hindu." Hosted in the United States, the site is targeted at a diasporic, primarily U.S.-based, Hindu customer base, as indicated by the facts that (1) the prices of products are in dollars and (2) some of the products, such as baseball caps, would have low market demand in India. Both these commercial Web sites reflect the marketing of Hinduism as a globally saleable product. They define the act of economic purchase as a cultural obligation to Hinduism. The activity of consumption, actively encouraged, is itself branded as Hindu. Participation in a chat room discussion about Hindu culture or history on Hindu Universe, followed by the act of purchasing a "Hindu" calling card or cell phone through the same portal, allows an economic transaction to be recast as part of a range of practices of cultural loyalty. Likewise, the Mobilehindu Web site (http://www.mobilehindu.com), which is also part of the Global Hindu Electronic Network, sells U.S.-based viewers cellular phone subscription packages from Nokia, Cingular, or Verizon, much like any other e-commerce gateway. However, as the title of the Web site suggests, the entire transaction is marked as an exchange between Hindus.

As in the Hindutva Web site, in sites linked through the Hindu Universe portal, the Indian state is subsumed under the broader category of an Indian, essentially Hindu, civilization with its roots in the mists of antiquity, all-knowing of scientific and spiritual truth. The Hindu Temples and Organizations page of the Hindu Universe site (http://www.hindunet.org/hindu_organizations) links to the Hindu Vivek Kendra Web site (http://www.hvk.org). On this site, a report titled "Hindu Agenda for the State of Bharat" avers that Bharat, or India, "has an antiquity untraceable by history" and has been "the soul of the universe radiating wisdom and knowledge—both of material sciences and spiritual experiences" (http://www.hvk.org/specialrepo/agenda/agendaenglish.html). The unique ethos of Indian civilization is described as *Sanatana Dharma*, which reflects Hindu ideals. The report asserts that the Indian state is obliged to build the nation on these scientific and spiritual ideals. The Sciences category of the Hindu Universe portal describes Vedic sciences, astrology, and expertise in astronomy and mathematics as part of the culture of ancient India. The History category of the site contains articles by Vartak (n.d., a, b) and Saraswat (n.d.) that claim to "scientifically" date and conclusively establish the historicity of events in the Hindu epics of the Mahabharata and Ramayana. And while the home page of the Hindu Universe portal does not appear to be prejudiced against Islam and Christianity, one has to but dig slightly deeper for the bias to become visible. For example, the content in the Interfaith Relations category on the site fulminates against Hindu converts to Christianity and Islam by force and deception. The rhetoric of Muslims and Christians as anti-Hindu and anti-Indian outsiders is also manifest in the History category of the portal and on linked sites such as Hindu Vivek Kendra.

In both Web resources, Hindutva and the Hindu Universe, Indian civilization is equated with Hindu civilization. In both Web resources, the claims of the nation supersede those of the state; in fact, a discourse of the state is nearly or completely absent from the Web sites. In both Web sites, the professed universalism of Hindu (which is treated as synonymous with Indian) culture does not extend to Indian minorities; on the contrary, they are marked by strong antiminority sentiment. While both Web sites stress scientificity, the Hindutva site emphatically stresses the scientificity of Hinduism as empirical and rationalist in nature. The Hindu Universe appears more accommodating of the claims of ancient Indian astrology and metaphysics as legitimate (Hindu) science. In contrast to the focus of the Hindutva Web site, the multipurpose Hindu Universe portal and related sites present a more conventional, if diffused, definition of Hindutva. The Hindutva site seems to amplify the idea of Hindutva as a logical, rationalist doctrine, whose value does not necessarily derive from any religious or spiritual significance. The Hindu Universe, on the other hand, does not shy away from highlighting Hindu religion and spirituality as central to Hindutva. The resources available and linked through the Hindu universe are also critical of secularism per se, unlike the Hindutva Web site, which defines its project as secular. Two examples of articles critical of secularism are the "The Intellectual Scene in Post-Independence India" and "Secularism: Indian Istyle," both found on the Hindu Vivek Kendra site, which is accessible through the Hindu Universe. The Hindutva Web site locates globality in a Hinduism garbed in a future-oriented discourse of secular rationality, while the Hindu Universe explicitly roots globality in spirituality and religion.

Conclusion: Rethinking Virtual Citizenship in the Age of Global Technology

The Hindu Universe Web site appears much closer to the Savarkariya model of the Indian nation-state than to the Nehruvian. However, the two general propositions found in both Savarkar and Nehru—the nation as cultural surplus over the state and the insistence on scientificity as a cultural attribute—are present on both sites. Similarly, the invocation of these propositions as the foundation of an imagined global Hindu-Indian nation is also common to both resources. In concluding this article, I briefly consider what these structural similarities and substantive differences between these Web resources might indicate about Hindu nationalism at large. I will also very briefly consider the implications of the virtuality of the online global Hindu-Indian nation, that is, what features of this imaginary might be foregrounded by the fact of its representation in cyberspace?

What is particularly interesting in the outline of the global Hindu nation, delineated in both the Web resources, is the contrast between two spatial-cultural relationships: (1) the relationship between the Indian nation and the globe and (2) the relationship between the Indian state and the globe. In these Web resources, a discourse of the nation is manifest, as is a discourse of the globe. What is nearly completely or totally absent, in contrast, is a discourse of the *state*. The reasons for the absence are not very difficult to fathom. Explicitly acknowledging a discourse of the state would require an engagement on the part of Hindu nationalist ideology with the *constitutional fact* of equal citizenship for members of all Indian religious and cultural communities. A mere acknowledgment of this fact profoundly complicates the Hindu nationalist argument. If Hindu nationalism rejected the constitutional fact of equal citizenship as incompatible with Hinduism or Hindu culture, it would undermine the claims of Hinduism as secular, inclusive, and universal. It would also undermine the claim of Hinduism or Hindutva as an eternally valid doctrine across time and place, since the rejection of equal citizenship would reveal Hindu culture as incapable of fulfilling the demands placed on it by the institutions of the *modern* state or the *Western* state. On the other hand, if Hindu nationalist discourse were to claim that the idea of equal citizenship is derived from Hindu culture or Hindutva, that would mean acknowledging that Hinduism grants full rights to all non-Hindu Indians. That, in turn, would contradict the Hindu nationalist argument that Muslims and Christians are outsiders in the Indian nation.

Second, an emphasis on the state and citizenship also problematizes the claim of expatriate or nonresident Indians to speak for the Indian nation. Given that a key center for the production and consumption of Hindu nationalist discourse is the diaspora, especially in the United States, it is likely that many of the visitors and users of Hindu nationalist sites are not Indian citizens. Accordingly, the claim to speak for the Indian nation must be located in a universalism and globalism that is simultaneously Hindu and Indian while it ignores the question of citizenship. The dual emphasis on the interconnected, globalized world of today and Hindu axioms about the supposed oneness of the human family are examples of such justificatory arguments offered by online Hindu nationalism. Importantly, in Hindu nationalist discourse, the universalism required by the nation and world in the age of globalized modernity must be *identical* with the universalism of Hinduism. A partial overlap, suggesting multiple notions of belonging that may not be exactly identical, will not do.

For that might open the door to the possibility that the universalist claims of various communities to inhabiting the Indian nation and the world are equally legitimate, no less so than Hinduism. The foundational status of Hinduism as the unique and ideal basis for national (Indian) and global (human) culture, accordingly, would be called into question. As another explanatory factor for the absence of the state, we may also note Chakravartty's (2001) observation that the "nationalism of the NRI is based upon a nationalist business culture that is pro-nation but anti-state," which is echoed in the mode in which commercial sites target diasporic Hindus as consumers (p. 346).

This preferential emphasis on the nation and globe over the state may also be found, to some extent, in offline Hindu nationalist discourse, whether diasporic or located in India. Hindu nationalist discourse within India would focus more on the state, although an emphasis on globality is very much part of the discourse, even if limited to rhetoric. As Hansen (1999), commenting on global aspirations of movements such as Hindu nationalism, has observed, "they want to arrive as national, sovereign modernities—as 'lights onto themselves'—and thus be recognized as respected members of that elusive global 'comity of nations' that remains the most sublime object of desire among even the most parochial nationalists anywhere" (p. 234).

What might be specific to its *online* manifestation is the *virtual construction of global nationhood*, enabled primarily through the interactive modes of Internet communication and its global, transnational reach. The Internet allows for a transnational group of users to engage in a variety of "Hindu" practices across national borders and time zones and thus to construct themselves as a global nation. This is strongly suggestive of Anderson's (1999) notion of the imagined community of the nation enabled by print capitalism, where the novel and the newspaper provided the "technical means for re-presenting the kind of imagined community that is the nation" (p. 25). However, there are some differences between the mode in which the global Hindu nation is constructed online and the modality of imagining the nation suggested by Anderson.

First, in contrast to the "passive" model of imagining oneself part of a nation through an act of reading books or newspapers, belonging to the global Hindu-Indian nation involves a profoundly performative commitment; loyalty to the nation must be expressed through combating anti-Hindu diatribe, purchasing Hindu items, sending Hindu e-cards, and the like in cyberspace. Rajagopal (2000) has pointed out the centrality of the Internet to Hindu nationalist discourse and the mode in which sympathizers of the ideology actively and creatively use the technology and medium in virtually imagining the Hindu nation into being. As Rajagopal notes:

> The internet is seen not simply as a channel of communication, but as a key battleground that must be occupied and defended as the always copious varieties of "pseudo-secular" criticism proliferate. In the words of the Nation of Hindutva, if the internet is "the last arena on earth where censors do not suppress facts," the advantage it allows suppressed facts can be more than made up by "the vast array of anti-Hindu and anti-Indian propaganda" circulating through the web. (p. 485)

Second, Anderson (1999) argues that "the nation is imagined as limited...No nation imagines itself coterminous with mankind" (p. 7). The global Hindu-Indian nation appears very much an exception in this regard, for it is precisely the claim of online Hindu nationalist discourse that Hinduism is identical with all things valuable that mankind has produced. Hinduism is the natural history of man. As Lal (2001) points out, the Web sites of "cyber-diasporic Hinduism" promise that the global spread of Hindutva will reinscribe the world as a global Hindu nation (p. 179). Toward this end, some of the Web sites abound in interpretations, as specious as innovative, of Argentina as "Arjuna Town" and Denmark as "Dhenu Marg" (Lal, 2001, p. 204)! Indeed, cyber-diasporic Hinduism often highlights relatively obscure elements of Hindu nationalist discourse, such as the VHP's "expansionist program for the Indian nation-state" (Lal, 2001, p. 215) to encompass the entire globe under the Hindu nation in the next century. Interestingly, this ambition resonates with Savarkar's (1969) claim in *Hindutva* that "[t]he only geographical limits of Hindutva are the limits of our earth!" (p. 119). As indicated earlier, the conflation of the Hindu nation with the globe takes the form of a spurious cosmopolitanism that garbs nationalistic concerns as global, universal matters. This ersatz universalism simultaneously seeks to define the national outsider, the Muslim or Christian, as unworthy of humanity itself, while Hindus, in contrast, are model members of the human race. The Hindu nation, at least in cyberspace, very much imagines itself as coterminous with mankind.

The third point of contrast with Anderson's (1999) argument is that the simultaneous, "homogeneous, empty time" of the imagined community of Anderson's nation is replaced by the eternal time of the global Hindu-Indian nation. The present is flattened out in a transcendentalized image of Hindu historical time. It is precisely this flattening out of time that enables the erasure of the specificity of the Indian state and statehood. If statehood is but a moment in the natural unfolding and flowering of Hindu civilization over time, then the obligations of this historical moment, such as the fact of equal citizenship, can be ignored by invoking both past and future. Although this possibility is not unique to the medium of the Internet, the ease with which alternate chronological narratives can be created in cyberspace may suggest a reason for the abundance of this account of Indian civilization on the Internet.

The mode in which online Hindu nationalism selectively draws on aspects of Nehru's thought is an example of how elements of progressive arguments can, in cyberspace, be decontextualized, coopted, and assimilated into a contrarian ideological perspective. While Savarkar's influence is clearly visible in online Hindu nationalism, his arguments are typically adapted to suit the present historical moment of global modernity. Cooption and reinvention are not unique to the Internet or to the Hindu nationalist project, but the possibilities of communication and representation offered by the technological medium seem especially apposite for both the substantive propositions—for instance, claims of globality and scientificity—and the tactical aims of Hindutva.

It is perhaps obvious to say that Hindu nationalism or, for that matter, diasporic Hindu nationalism is not a monolithic entity. Mathew and Prashad (2000), for instance, have described the "protean forms" in which "Yankee Hindutva" reflects its various motivations and seeks to materialize its objectives. Online Hindu nationalism, likewise, is forged in the context of multiple commitments, even as it will use whatever is grist to its mill. The framework of globalized modernity, within which Hindu nationalist Web sites are produced and consumed, influences Hindu nationalist discourse such that the latter draws, in instrumental and piecemeal fashion, on multiple and contradictory legacies of discursivity toward accommodating the demands of global capitalism. In return, online Hindu nationalist discourse mobilizes the resources, economic and technological, of global modernity in articulating different logics of persuasion to propagate itself. One hopes that this symbiotic relationship does not reflect an especially virulent strain of the ideology.

CHAPTER TEN

The Hybrid Cultures of Cyborg Diasporas: Making Sense of the Expatriate Odias' Conversations

Anustup Nayak and Natalia Rybas

Can You Be Your Own Self in Cyberspace?

MANY YEARS AGO, *The New Yorker* ran a cartoon of a dog sitting in front of a computer, gleefully logged on to some virtual community of humans. The caption read, "On the Internet, nobody knows you are a dog!" That dog became an enduring symbol of the very essence of cyberspace. If the story of the dog was to be believed, cyberspace was destined to liberate its denizens from every conceivable barrier imposed by real life—race, location, gender, class, nationality, and even species. As global flows of technology, human beings, and culture begin to compress space and time, many assume that we will stake out new frontiers of imagination, connection, and creation. Liberated from the constraints of space and time, disconnected from history and politics, and united by interests and purpose, cybernauts will lead the project of creating compassion, commonality, and community. At least that is what the pioneers proclaimed about the advent of the Internet. However, the contesting voices of the members of the online community of the Odia-Indian network called Ornet reveal how the baggage of cultural, political, and psychological realities of India, especially the state of Odisa, can complicate this supposedly liberating project of the virtual community. As one member writes,

> That writing opened up the door for establishing connections with my missing friends. Mami [Minneapolis, MN, now in Atlanta, GA] wrote to me "Bigyani Nani, is it you?" Arati Nanda from Austria [now in Houston, TX] wrote, "Bigyani Apa, it's you I am sure." Pada [Padmanava Pradhan] from Sweden [now in Charlottesville, VA] wrote, "Bigyani, I think it's you." Bandana Apa from Australia wrote, "Bigyani, I think you are the same Bigyani, Krishna's friend." And it opened the door for many of my friends.

"Apa" and "Nani" are Odia terms for elder sister. In Odisa, as in much of India, people refer to older social contacts affectionately as Apa (sister), Bhai (brother), Mausa (uncle), and Mausi (aunt). These appellations are at the heart of conversations and relationships in many Indian cultures. Using different terms ("sister" vs. "aunt") may actually indicate the relative age difference of the other person. Calling a stranger Apa or Bhai is the essential step to relationship building in Odisa. This quote suggests that Ornet is not a just a virtual dungeon of anonymous beings whose identity is entirely the creation of the computer screen. It is rather the projection of incredibly real and strong personal ties, forged in the schools, homes, and marriages of far-away Odisa, recreated and transformed through immigrants and technology.

Identity of person, family, and society is at the very essence of Ornet. These voices challenge the assumption that technology and globalization dissolve age-old markers of human existence. The upheavals in Odisa over several hundred years color the tensions, feelings, and purposes behind the several hundred email messages floating through the culturally circumscribed cyberspace of Ornet. Members are articulate and contentious; their words open every single linguistic, cultural, and political fracture there is. They accuse and argue, concede and compromise, prefacing their tirades by affectionate traditional Odia appellations of Bhai, Apa, and Mausa, or sometimes in the most vitriolic swear words of Odisa's streets. Ornet's voices are a bundle of contradictions, not easily explained by existing renditions of cyberspatial dialogues, much less by explanations of postcolonial immigrants' angst.

This chapter explores the conversations archived in Columbia University's computer systems. The narratives and dialogues of the Ornet system touch upon questions of cultural heritage, connections between the old home and the new home, national character, and the relationship between the mother tongue and the English language. The community of immigrants on the Internet creates a cyborg diaspora where the members negotiate who they are and who they become in the geopolitical structure of the globalized world.

The Ornet Virtual Community: A Formal Introduction

Ornet is an online community consisting of immigrant Indians from the eastern state of Odisa who have moved to the United States. Odisa, their home, is a small state on the eastern coast of India, with a thousand-year heritage of language, art, and culture. In recent decades, Odisa has been marginalized by poverty, corruption, and natural disasters, and now it is becoming one of the most impoverished places in India and perhaps in the world. Ornet mostly consists of Odias staying in the United States, but some are back in Odisa, and a few are in other parts of the world. A few Odia graduate students in the United States launched the Ornet listserv and the accompanying Web site, the Orissa homepage, in 1990. The project started almost by accident when Subrat Mohapatra of Stillwater, Oklahoma, posted an announcement on the soc.culture.indian newsgroup asking whether people wanted to accompany him to the Odisa Society of the Americas (OSA) convention in Chicago. The forty people who responded became the core of a network that has grown to about a thousand people from all over the world today. The original founders, Subrat Mohapatra, Chitta Baral, and Ashutosh Dutta, are still active on the list. Ashutosh Dutta maintains the list from a server in Columbia University (Dutta, personal communication, January 15, 2002). Gradually, more people joined the list, and it became the center of dialogues between the members of the Odia diaspora and the online forum for many Odia social organizations that grew from the immigrants' efforts. Other forums for Odias to engage online, such as the Odisa list and the Calnet list, have emerged in recent years; many started with the help of former Ornet members.

One of the chapter's authors, Anustup Nayak, narrates further. My interest in Ornet is deeply personal. I have been intimately connected to Ornet and its members. I was born and raised in Odisa, perhaps within only a few hundred miles of the homes of all Ornetters. I studied with some of them, met many in social gatherings, and became related to a few others through friendship and family ties. On the one hand, I loved Ornet. As a graduate student in Georgia Tech, I was new to the world of the Internet, and I still am to the alien surroundings of America, and Ornet offered amniotic safety in the voices of fellow Odias. It offered assurances, connections, and memories of a home left far behind. Many members were eminent Odia intellectuals whose names I had only heard in childhood and whom I was happy to correspond with. On the other hand, I hated Ornet. Relentless wrangling over seemingly petty issues of language and culture, the one-upmanship over agendas and actions, the incoherence and the sheer deluge of emails sometimes left me exasperated. After two years of active participation I left Ornet in 1998. I rejoined it again in 1999 in the wake of the vicious cyclone that claimed many thousands of lives in Orissa. Ornet members created a wonderful linkage of hearts, minds, wallets, and hands to revive the homeland torn to shreds by the gales and waves. I left again, only to come back as a "lurker" trying to understand this space in the context of the broader forces of technology, globalization, and culture that consume my time these days. Ornet had changed substantially, focusing on a few specific linguistic and political issues. Many members have left, many new voices are there, and some of the older ones are still as vocal.

Communications of transnational diasporas and online virtual communities have been extensively researched, albeit as two disparate academic exercises. There is potential to combine these two methodological lenses. The Ornet community reveals some features that are not as conspicuous in the context of anonymous and purely online interactions of mostly Western individuals. As I read through the transcripts of conversations on Ornet, many questions come to mind. Does Ornet have mostly immigrant narratives, or is it akin to other virtual communities? How are offline transactions between members and Indian community norms relevant in these discourses? Are Ornetters recreating in cyberspace a sense of what "home" is? How do the pressures of Odisa's past, present, and future mediate these discussions? What are the political, psychological, and cultural drivers of such a reconceptualization? How does Ornet play a role in the development of civic engagement of this immigrant community among its own members and with social progress in Odisa? There are no definitive answers, but this chapter sheds some light on our understanding of the technology-mediated interplay among nationalism, culture, and community.

A combination of the archives of Ornet; email, phone, and in-person interviews with members; and news articles about Odisa serve as sources for the analysis. The names of the members have been omitted in order protect their identities. The reading of the voluminous emails is more selective rather than comprehensive, concentrating on those discussion threads that reflect major points of fracture or conciliation in the community.

Cyborg Diasporas

Today, thousands of people move voluntarily and involuntarily to seek economic and political opportunity across geographical boundaries, redefining the basis of ethnicity, culture, and nationalism. Travel between their two or more homelands, new community organizations, and political coalitions facilitate the processes of cultural assimilation, hybridization, and contestation that immigrants experience. New media, including the Internet, emerge as a very important platform for the cultural and personal experiences of immigrants, especially those who settle in technologically advanced Western countries. Through email listservs, chat rooms, and Web portals immigrants create a rich set of interpersonal and intersocietal ties.

Traditional explanations of technology-mediated globalization paint a picture of a world where the borders of the nation-state, elements of traditional culture, and pathways of history are readily dissolved by advances in information and communication technology. Information technologies also become the substrate on which other social formations, commonly referred to as virtual communities and communities of interest, can take shape. Such communities were predicted by Licklider, the creator of the first computer network, ARPANET, as early as 1968 (Rheingold, 2000). Licklider asks, "What will online interactive communities be like? Life will be happier for the online individual because people with whom one interacts most strongly will be selected more by commonality of interests and goals than by accidents of proximity" (p. 9). Rheingold recounts his experiences trying to actualize the vision that Licklider presented almost thirty years ago. Rheingold re-reiterates the dissolution of physical geography and personal identity when one participates in an online community. In Rheingold's world, where the human physique is hidden behind the simulated persona of the computer screen, identities of individuals are fundamentally transformed and they become primarily "transmitters of ideas and feeling beings, not carnal vessels" (p. 11). Individuals begin to assert more power in their social interactions, actively choosing others with common interests, using powerful multicriteria search engines that are able to index identities based on attributes that others choose to reveal about themselves. People choose whom to connect with, when to connect, and what to connect about. Creativity, independence, and even consensual interdependence become the hallmark of such virtual communities.

Rheingold's (2000) world is not too far from coming true. Today, literally thousands of communities of interest litter the wilderness of the World Wide Web, connecting individuals, far apart in location, culture, and personality, yet united by common interests, for example, in supporting a political cause, such as abortion rights, or providing emotional care for common medical conditions. However, Rheingold's world is not entirely true either. Technological connectivity is increasingly blamed for the isolation, disconnectedness, and individualism that pervade the social sphere in many communities today. While making it possible to reach out to others across the barriers of geography and culture, the twin forces of communication technology and televised entertainment have made people increasingly isolated and self-centered. The soothing comfort of speaking, touching, and sensing—the very basis of togetherness—is being increasingly replaced by the impersonal stimuli of voicemail and email. Putnam (2000), an influential American political scientist, notes that Americans are increasingly disconnected from their families, neighbors, communities, and the republic itself. Putnam argues that one of the biggest culprits for the demise of public engagement is the national obsession with television and its synthetic imagery, which encourage people to be cloistered in their private spaces. While technology certainly provides the infrastructure to build networks and communities, it may not provide the social and psychological impetus for such engagement, and if the skeptics are correct, it may indeed serve to inhibit it.

The emergence of Internet networks of immigrants provides a complex counterpoint to such divergent understandings of technology-mediated communities. "Cyborg diaspora" as a term was originally conceived by Karamcheti (1992), much before the proliferation of the Internet, to talk of a possible technology-enabled community that would help disrupt the fossilized notion of "Indianness" within Indian immigrant communities. This conception has been actualized in many online communities created on the Internet by immigrants. Gajjala (2004), another Indian scholar, rejuvenates Karamcheti's cyborg diasporas to "refer to the formation of virtual imagined communities of diasporic post-colonials online" (p. 13). The expressions of immigrants have been characterized by the conception of an "imagined community," an extension of the boundaries of the nation-state and geocentric ethnicity created by the imagination of people separated by distance yet united by the twin forces of print media and capitalism.

> The nation is an imagined political community because the members of even the smallest nation will never know most of their fellow-members, meet them, or even hear of them, yet in the minds of each lives the image of their communion. (Anderson, 1991, p. 6)

In recent years, the Internet has usurped the role of print capitalism as the master mediator of the imagination of "mother country" and the "adopted country." Virtual networks have become sources of news, forums for social and political engagement, and sites of community building. This project of redefining a nation in a virtual space has been described as "taking a nation-state completely out of its traditional context, material space enclosed by politically drawn boundaries, locality, and reconstructing it or some simulacrum of it inside an immaterial or virtual space" (Bastian, 1999, Introduction). Such spaces lie at the intersection of the geographically determined communities that exist in the real world and the communities of interest imagined by Rheingold (2000). They also lie in between the democratizing and liberating promise of virtual communion and the replay of the contentions, fragmentation, and power structure that exist in the real world as though an alternative type of social space within which virtuality and ethnicity are fused (Da Rosa, Gold & Lamy, 2000, p. 5).

The dynamic of the so-called cyborg diasporas is not limited to the tension between the centrifugal force of virtuality and the centripetal force of ethnicity. It has also become an important connection between the political, economic, and cultural lives of immigrants in their home and adopted countries. This projection is achieved by the intimate relationships between virtual communities, such as Ornet, and real-world immigrant organizations, such as the Odisa Society of the Americas (OSA), and the complex chains of influence between the two. The boundaries between disparate but "real-world" immigrant experiences, such as weekend get-togethers of Odias in a Washington DC suburb, prayers at the Jagannath temple in Nashville, and postal ballots for the election of OSA officers, and events in India, such as legislative decisions of the state government in Odisa, are seemingly glued together through emails floating in cyberspace.

This project visualizes the conversations on this virtual space as embedded in the discourse around cultural globalization yet distinct from the standard Westernized renditions of its two components—virtuality and ethnicity. Such an approach helps us conceptualize globalization not as an all-changing zero-sum tug-of-war between distinct forces of technology, geography, and identity but as a hybridized construct of incremental changes in interpersonal and societal relationships. The value of such studies is not merely in reconsidering existing beliefs but in reconstructing the basis of community in a complex world. Ornet and similar spaces are the locations where the borders between the real and virtual worlds are increasingly amorphous, time and space are constantly reconfigured, and imagination is inextricable from action. These are the sites where the community of the future will take root.

Implications of Odisa's History on Ornet

When members post their thoughts to Ornet, they voice their hopes, ideas, and frustrations about their home state, they remember times gone by and places lived in, they forward both sensational and insignificant news items, and they write eloquent poetry and sometimes crude jokes. In these acts they not only recreate their own selves, in ways only the Internet lets them, but become intimately intertwined with the history, geography, politics, and culture of Odisa. Odisa is a place of contrasts. A small state situated on the shores of the Bay of Bengal and stretching into the mountainous heartland of India, Odisa is a microcosm of the diversity of India, reflecting many cultures, languages, and religions. Its rich heritage, traceable to the sixth century BC, is reflected in the architecture of the magnificent temples at Puri and Konarka, in the colorful designs of Sambalpur and Berhampur saris, in the ancient traditions of the Kondh and Santal tribes, and in the exquisite postures of the classical Odisi dance. In addition, the story of Odisa is one of poverty, starvation, exploitation, and decay. Two-thirds of its rural population live in abject poverty. It has India's highest infant mortality rate, the lowest number of doctors per capita, and one of the worst records in the country for providing electricity and water to its people (Lak, 1999). Floods, cyclones, disease vectors, colonial exploitation, and postcolonial apathy from the Indian federal government dealt Odisa a very bad hand in the poker game of economic development. At the same time, corrupt politicians, inept bureaucrats, an apathetic elitist class, and the lack of entrepreneurial action played that bad hand extremely poorly.

The virtual space of Ornet, born from the cultural tissues of Odisa, is riven with the same angst of a glorious past hidden in the morass of an insecure present. One member posted,

> Unique seal pendant discovered in Orissa (United News of India) Bhubaneswar: Two archaeologists have discovered a unique seal matrix engraved on a pendant made of red jasper from an early historic site of Budhigarh in western Orissa ... The artifactual evidence from the site indicated that it was occupied from the early historic to the medieval period. The site has already revealed a rich ceramic industry, terracota, enormous quantity of cast and punched marked coins, iron implements, faunal remains, brick structure and lavish use of beads made of precious and semi-precious stones.

Another post noted in sharp contrast, a few days later,

> Dear all, while we are dealing with peripheral issues and sniping at each other, Orissa is sinking by the day. Here is a report in today's Dharitri: About 90 percent people under BPL [below poverty line] category are homeless: WDR Bhubaneswar: With the release of the World Development Report-2001, the details regarding the poverty-stricken people in the state has come to light. The report has unveiled the life style of paupers, landless and homeless people in the state.

Here the members of the online community celebrate the glory of their cultural heritage and lament the uncertainties and disparities of the present. However, for many members, especially those educated in economics and the sciences and liberated by the intellectual freedom in the United States, the future is a place to be as it promises hope. Another post states,

> Poverty of a given individual, community, or state is a "vicious cycle." We need to identify the weakest link in the cycle, break it and linearize the problem. Once the problem is linearized, we now know the beginning and end. Then one/all can work incrementally to shorten the length of the problem ... A good thing about poverty is that when we are at base 0, anything we do is a progress. Therefore both visible progress and visibility of progress is easy to accomplish in Orissa. Thus, Orissa is full of opportunities and has unlimited potential for individual contribution; a Gujurati may not have such flexibility. Let us try as much as we can.

The connections between the conflicts in Odisa and their representation in the messages posted to Ornet are rather complex. On the one hand, they are the constructs of the imaginations and personal experiences of the writers. Visions of Odisa are often fossilized in the moment that the diasporics first boarded the flight to come to the United States. For some senior members of the diaspora, it is as long as thirty years ago, when the current capital city Bhubaneswar barely existed as a few colonies of government housing. For some others, who arrived in the United States very recently, the sights and smells of home are rather different. The waves of Odia immigrants to the United States, though no accurate record has been kept, can be roughly translated to three phases—a few pioneers in the 1960s, increasing in the 1970s and 1980s with the drive for scientists and engineers during the cold war, and then the massive influx of immigrants during the software boom of the 1990s. Ornet's voices are thus sharply divided between the "old" and the "new" arrivals, one reminiscing about an age gone by and another struggling to make sense of a world rendered empty by broken promises of the past generation. While this "old guard" claims connections to a golden era of eminent Odia freedom fighters, eloquent poets, and a blossoming language, the "newbies" reveal both the frustration of growing up in corruption and underdevelopment and the detachment of modern anglicized culture from the heartland of Odisa. The clash of generations is often intense, yet the gentle family-like persona of Ornet creates a space where differences are mediated through highly personal interactions, sometimes through offline phone calls, and even through visits to each other's houses.

The connections between Odisa and Ornet are also reinforced through what can best be described as "bidirectional social remittances" of information, culture, and politics that travel between Odisa and the United States. The term "social remittances" was coined by Levitt (2001) to describe the "ideas, behaviors, identities, and social capital flowing from the host to sending country communities" (p. 11). Levitt asserts that three forms of remittances occur: normative structures, such as ideas, values, and beliefs; systems of practice, such as socializing and participating in civic life; and forms of social capital, such as prestige and status. All of these three forces flow from Odisa to the United States when people call home every weekend, fly back for a few weeks on Christmas, or receive an occasional guest from back home. These small remittances add up and spill over into the online discussions, coloring the words and voices of members. However important the effects of these remittances may be in terms of building the online discussions, these exchanges in themselves represent a somewhat superficial connection with Odisa's reality. A small number of nonresident Odias, identified with an urban middle class in Odisa, relatively disengaged from civic life, and even more distanced from the poverty pervading the rural heartland, make the voices of Ornetters somewhat paternalistic, disengaged, and hypocritical. As Bijoy Misra, a Harvard academic and senior member of Ornet, notes, "In post-colonial Odisa, a sense of survivalism has taken up a large measure of the lives of people and the 'soil' is forgotten." This disconnect from reality results in a fetishized version of the homeland, where all that remains are images either of extreme glory or of extreme despair. Tired of such extremist interpretations, disgruntled Ornet members tend to exercise "reform through exit," creating alternative email and local forums, diminishing both the number and the diversity of opinions on Ornet.

At any point of time, only about twenty members are active in discussions on Ornet; the remainder passively lurk. Few women join the discussion, but the ones who do are extremely vocal and articulate. There is a sense of hierarchy among the members, reflected in age, level of participation in other Odia cultural activities, the intellectual nature of their professions, and, of course, social status in Odisa. There are also a few members who have built their reputations (and notoriety) through their participation in Ornet. Some members have taken on the role of mediators between parties, others have acquired identities on the basis of their ideologies (for example, the member who speaks of the Killer Instinct or the member who advocates spirituality), and some others have been rebuked as irritants. This skewness in voice and participation is characteristic of public life in Odisa, where social hierarchies strongly determine public participation.

Voices in Odisa's public life are thus split between corrupt demagogues and idealistic intellectuals, all far removed from the daily travails of the majority. The garrulousness of some members and their propensity to offer opinions on all imaginable subjects are representative of such characteristic of an elitist civil society, whose small numbers and concentrated social influence tend to make them "know-it-alls" of public life. Thus many members find it both difficult and distasteful to engage in Ornet, as they do in Odisa's public forums.

Despite their perceived differences, weaknesses, and hypocrisy, Ornetters have successfully tried to strengthen the flow of social remittances back into Odisa. Many organizations that engage in social and economic development are tied to Ornet using its space to mobilize membership, create political consensus, organize gatherings, and also request support for individual projects in Odisa. In the wake of the cyclones and floods that have ravaged Odisa in the past few years, Ornet has become a tool to raise funds and monitor the progress of projects to rehabilitate affected communities. As eminent scientists, academics, and businesspeople, Ornet members exert a small but increasing influence on the political and bureaucratic processes of Odisa. Ornet constantly gets involved in protests, petitions, and lobbying Odisa's politicians and bureaucrats. Members of Ornet have gone on to create organizations such as the Sustainable Economic and Educational Development Society (SEEDS) and the National Conference on Information Technology (NCIT) that support social entrepreneurship and information technology research in Odisa. The very frustration of not being able to achieve concrete things inside of Ornet has given these committed individuals energy to pursue their own dreams outside the virtual diaspora to make changes in real life.

Cultural Contestation through Hypertext Interactivity

The role of the culture, values, attitudes, norms, and beliefs of people in the economic trajectories of nations and societies is one of the hottest debates of economic development. Whether culture is the sole determinant of economic progress or a mere proximate symptom of deeper geographical or historical factors becomes the common starting point for diasporics when they envision the future of their homeland. For societies, such as Odisa, that have seen their fortunes falter in sharp contrast with their ethnic neighbors, the role of culture and values is the most torturous truth of being. The interactivity and the distance provided by the virtual space enable the discontents of Odia culture effectively to question some of the deeply held attributes of Odia culture in the light of Odisa's underdevelopment. One member stated raucously when he wrote,

> it is frequently brought out to attention that we odias have less "killer instinct." it is said that because of lack of this we are not as successful as we could be, i.e. we fail to attain our achievable potential. this observation is nothing new but something a few of us have been harping along for over 2 years. equivalent terms denoting this attitude are "pakhala khai soi pada" [eat watered rice and sleep], "alpake santussta" [content with minimality], adaptable, peace loving, positive, fatalism, "kapale likhana ke kariba kana" [you cannot erase the fate lines on your forehead]. pacifism, peace loving, positivism everything are all very good ideas provided it has the "killer instinct" to drive their objectives. else all the hoopla about positivism is nothing more than a charade to gloss over one's inadequacies. nothing wrong with that, except that it prevents a forward push.

His perceptions are indeed shared by many Odias in and outside Ornet, and perhaps even by many outsiders who perceive an Odia as a peace-loving, shy, and laid-back person, not very keen to "push forward." To these observers, it is no surprise that corruption, inefficiency, and misrule went almost uncontested and that underdevelopment was the end result.

Taking exception to this line of thinking, one could argue that "killer instinct," or aggressive and progressive values, may not be entirely alien to those born in Odisa. Odia fighters under Surendra Sai fought the British valiantly and were the last free fort to fall; a war hero, Subhas Bose, born in Odisa, aligned with the Japanese to launch an invasion of British India. And, if the ferocity of Ornetters' arguments is anything to go by, not all Odias lack the so-called killer instinct, at least in its verbal sense. There is also a second level of contestation to the killer instinct metaphor. Many believe that pacifism and fatalism are indeed the exquisite essence of Odia culture. The emperor Ashoka, who attacked Odisa and killed thousands of enemy soldiers, was transformed by the experience and became a pacifist Buddhist. Great Odia leaders such as Gopabandhu Das, Madhusudan Das, and Godavrish Mishra played prominent roles in India's nonviolent freedom movement. To give up these values is to disintegrate the very core of the national identity. To others, the issues of economic development, education, and civic engagement need to be solved before defining national character. In particular,

> Now the question is: Why Oriya culture or Oriya language is not prominently displayed like Bengali or Tamil or Punjabi? It has nothing to do with a killer instinct or a lack there of among the Oriyas. They are as good, bad, intelligent, brave and endowed with instincts as any other group in India. The fact remains that economically Orissa is still behind. There have not been enough of the technological developments in the state. Ordinary people, the backbone of a culture or a language still languish in Orissa in a great deal of poverty, illiteracy and apathy ... Sorry if it hurts, but I am ashamed of my friends who have nothing more than a self denigrating attitude towards their own kind.

Some others feel that breaking the chicken and egg conundrum—whether the character or the problems came first—is less relevant than using the online forum as a tool of economic development so that the deficiencies both of culture and of structural progress are concomitant. The very fact that volunteers living far apart in America and India could organize themselves to create socially relevant projects through collaboration enabled by the Internet gives enough killer instinct that can act as an engine for progress without denigrating the depth of language and culture.

Watching this debate unfold, I learned a lot about the complex relationship between culture and development. But what I really noticed was the effect of the digital medium on the nature and content of this debate, which plays itself out over and over again in public spaces in Odisa and the rest of India. Such virtual presence provided the safe distance from which the participating individuals could challenge their social superiors (mostly older persons in the community) with the most direct and the harshest of criticisms. It also enabled each respondent to bring in facts and figures, anecdotes, and examples in a way that is not often possible when these people sit face to face feeling the pressure of attention. In fact, I have watched these same individuals in Odia cultural gatherings, debating similar issues—loudly, monotonously, and sometimes without much reason. Even though Ornet is not without its share of flame wars, anonymous emails attacking members personally, and innumerable squabbles over seemingly minor things, the ability to engage in interactive dialogues and to ratiocinate with the willingness to concede and negotiate was quite pronounced. The asynchronous nature of Internet communications gives time to reflect. The archival capacity of email groups allows one to provide specific references to others' comments, creating context. The invisibility of the other imbues them with a certain courage, but the strength and vitality of social relationships among the members prevent that courage from being recklessly misused.

Searching for Roots: Can Technology Rejuvenate Traditions?

The debate on killer instinct did not come to any specific conclusion. Very few debates on Ornet do. Rather, the discussion metamorphosed into another debate on recreating the need and value for preserving the worth of the Odia language, an issue that has remained a bone of contention among Odias living in Odisa and abroad. The members realized that the lack of killer instinct and the course of economic development could not be changed overnight. What was indeed the cause (*karana*) leading to the effect (*lakhyana*) of underdevelopment was the failure to realize and preserve the relevance of and attachment to Odia language and traditions. Over the years, as Odisa fell behind other regions in economic development, the value of learning Odia and asserting Odia traditions became less pronounced among the socially mobile urban middle class. Many argue that it became almost a necessity to delink oneself from the so-called backward Odia identity and assert a more pan-Indian persona. As people moved out from

Odisa to live in other parts of India and in other countries, it became difficult to maintain Odia traditions outside the community, much less to pass them on to one's children. One member stirred up a hornet's nest by asking,

> Friends, I was wondering about this question: What is the definition of Mother Tongue? I may be wrong. But as I have understood, Mother Tongue is the language you speak all the time, you communicate with for all your needs, you think, you argue, you express your emotions with and it's the language of the land. When I was in Odisha, Odia was my mother tongue. I expressed everything in Odia, I asked my mother "bhoka helani, khaibaku de" ... Here everyday, for about 10 hours of my working life and at home with children while checking homeworks I communicate everything without Odia and using only English. I ask my friend, "I am hungry, let's go to a Chinese Restaurant" ... so if Odisha is my mother "Debaki," at present USA is my mother "Yasoda." I have my duties for both of these mothers and accepting this truth and balancing the duties are the right things to do as regards my opinion.

Her comments struck a nerve with many members, some of whom loudly protested her supposed lack of courage in accepting her "Odia-ness." One said,

> The explanation is not new and it is the widely used way to hide yourself from you. Even though you are trying to do something for odia language but from your statement it looks like you pretend to be an odia but you don't want to be an odia by heart.

In the same vein another commented,

> Only reason could be you are trying to invent a safe excuse for your future necessities like educating your children. There is nothing wrong in doing that as long as you don't build this excuse. Boldly express yourself: "Given the hectics of life in the US I am no more able to cope with the demand of effort that is necessary to live up to my Oriya roots."

Such vicious wrangling over the pragmatic realities in an anglicized global world versus the moral or cultural responsibility to adhere to century-long traditions are nothing new to immigrant communities. Many members on Ornet have lamented the use of English in Odia social gatherings. To them, speaking Odia words, wearing traditional dress, and eating home-cooked Odia food are the fundamental points of association with culture and heritage.

Some others have countered that mere use of a certain language should not be looked upon as an act of cultural desecration. To draw attention to the issue, a member started writing with Odia words embedded inside English sentences. To some it was quite amusing and to others quite confusing. One example is a sentence that runs like this:

> Everything seems to be so "neeraba, nischala" [silent stillness] in the odia e-front. It is said that odias can never be woken up if they have had some "pakhala" and "tentuli" water [favorite dish in Odisa is watered rice with a dash of imli—a fruit used to add sourness to a dish].

Here the speaker makes specific cultural references in Odia; even though the major fabric of the language remains anglicized, the insertions of mother tongue carry cultural significance and signify group belonging. Such references make sense only to the members of Odia diaspora, especially if they have certain baggage of history and culture.

Over the months, one Ornet member started postings called the "Odia word and thought of the month," which became an interesting project in resolving this zero-sum game between adopting the Odia or English language. Members would come up with an Odia word or a famous Odia saying that carried cultural significance and then try to interpret its meaning in English. In the process, the discussion participants created some form of Odia dictionary that members kept adding to through their personal knowledge and research. Each month the responsibility of coming up with the Odia thought of the month was taken up by a volunteer. Literal translations led people to explore the etymology of words and phrases. This spurred readers to come up with cultural and psychological connotations of word usage. One particularly avid participant dissected the Odia synonyms for "relationship"—*samparka* and *sambandha*:

> Sambandha is more immediate, someone you are "bound" with. All people in a family have sambandha. Sambandha has the connotation of mutual responsibility: sam+bandha etc. We have met sam before, a

> preposition "upasarga" meaning "well." bandh [is a] verb [that] stands for "to bind." samparka is distant and is based on respect and friendship. The root verb is ark, meaning to praise. The exact meaning will be one who is immersed in praise. Unlike sambandha which binds people to one entity, samparka retains the entities and has a bond of respect.

The "Odia word and thought of the month" did not take the steam out of the debate on the traditionalism versus modernism. Rather, it reordered the tensions of that debate into the energy of community activity. People began to participate in a process through which they co-created a language lost in the crevices of their memory, established the links of language with mythology and popular culture, and, most important, recalled the sights and smells of their youth. Odia words that once stood incongruously in the English sentences of the aforementioned contributor began to take on a life of their own. People started exploring the nuances of the language, which varied even across the small geographical spread of Odisa. One member prepared an extensive list of the names of family relationships and their variants across the different Odia dialects and across different castes. For example, uncle, or father's brother, would sound like dadei, badakhabapa, or kaka. Aunty, or father's brother's wife, would be dethei, khudi badakhama, or kaki.

For many young Odias living outside of the state, adhering to Odia words and customs has often been the result of coercion by elders and peers, an act born out of the fear of being delegitimized by one's community, an inconvenience forced on them by an accident of birth in a deprived land. For many years, the so-called protectors of Odisa's heritage, intellectuals, and politicians have been perpetuating this guilty feeling of "Odianess"—a fetishized nationalism that promotes adherence to a predefined package of beliefs, language, and customs. The adaptive process of wrestling through these issues on Ornet indeed provides a way out of this prepackaged Odia nationalism that often ends in rejection. Through their long-drawn collaborative work, these individuals have created a space where reconstructing linguistic order and accepting hybridized cultural practices become, for lack of a better description, a fun thing to do. When traditional values are associated with communal guilt, false nationhood, or axiomatic norms, tradition becomes a burden that people itch to offload. When presented as a privilege, as an opportunity, as an act of value creation, traditions assume a life of their own, even in the face of hegemonic cultural and economic countercurrents.

The Paradox and Promise of Community in an Interconnected World

I started my exploration with some conceptual descriptors to categorize the interactions that I observed on Ornet. I asked myself many questions: Is Ornet a virtual community, in the sense that it connects people who live far away and are unknown to each other? Sure it is. But do the Ornet members create a completely imaginary world that removes all barriers that their ethnic, geographical, and historical antecedents impose on them? Certainly not. Is Ornet a pleasurable experience? Yes, indeed. It brings to life relationships and memories of a lost homeland and acts as a platform of social engagement. It creates the space for rejuvenating a defeated psyche and a lost language. Is Ornet distasteful? Perhaps, yes. The personal attacks, the constant jockeying for attention, the fixation on symbols, the xenophobia, and the lack of consensus are indeed deplorable. Is Ornet a product of technology, ethnic identity, or global integration? The answer is that it is all of the above and more. Most of my questions have been answered with a set of contradictions. It is this set of inexplicable contradictions between personal and societal values, together with the complex interplay of technology and community, of transnational and local spaces, that make Ornet an engaging and valuable object of study.

These contradictions suggest the value of studying such projects, which show intersections of many forces of globalization: technological change, ethnic migration, and cultural harmonization. It is the tendency of intellectuals studying globalization to characterize the state of the world into a set of mutually exclusive realities, such as traditionalism versus modernity, real versus virtual, and ethnic versus transnational. They also tend to speak about globalization in terms of macroscale phenomena, tsunami-like transformations that go unrestricted across societies, economies, and nation-states, unless challenged by similarly large-scale forces. Many social and critical theorists use a so-called millennial framework while studying global phenomena such as the Internet (Sterne, 1999). In these scenarios, the critic considers the "impact" of "new" phenomena such as the Internet or global capitalism that "transform" the structure of the world in one way or the other. These critics engage in high-stakes intellectual debate that requires them to delineate narrowly the objects of their study. Researchers mistakenly ascribe autonomy to these defined objects by failing to associate them with other social forces and by remaining blind to their incremental impacts. Marx (1999), a technology historian, questions the assumptions and conclusions that stem from misinterpretation of technology's historical effect on economies. He says,

> It [technology] is depicted as if it were a palpable thing—a complicated kind of tool devised by a few geniuses, members of technically gifted elite—an embodiment of virtual autonomy and agency. "Technology" is tacitly accorded autonomous power in that its efficacy is perceived largely unconstrained by the circumstances of its origin or dissemination, and it often is implicitly endowed with agency, as if it is inherently capable of making things happened—of determining the tenor of human behavior and indeed, the course of human events. (pp. 136–137)

The existence of spaces such as Ornet and their complexity challenges these simplistic and all-encompassing visions of technology and globalization. Events in such virtual communities happen in close causal connection to the cultural forces of the physical world, and their impacts are neither cataclysmic nor unidirectional.

Although the impacts of the interactions on Ornet and other cyborg diasporas are neither dramatic nor univocal, they do represent a broader trend in the confluence of the forces of media and migration as a source of rupture in the existing social order (Appadurai, 1996). As people move across the boundaries of nation-states or other smaller geographical formations, in Appadurai's words, "these diasporas bring the force of imagination as both memory and desire, into the lives of many ordinary people"

(p. 6). Transnational mass media, and now cyberspace, deeply affect the politics and the process of adaptation of migrants as they move. The imaginations of home and abroad and the contentions over language and traditions on Ornet create the cultural norms and interpretive frameworks through which new Odia immigrants will view their new homes in America. Appadurai completes his formulation by saying,

> The theory of the media as the opium of the people needs to be regarded with great skepticism ... where there is consumption there is pleasure and where there is pleasure there is agency. Fantasy can dissipate (because its logic is often so autotelic), but the imagination especially when collective, can become the fuel for action. (p. 7)

The ability of the Internet to personalize the creation and consumption of information has a great potential to create a resistive force against cultural domination. The diversity of forms and content of communications, made possible through the interoperability standards of the Net, has a potential to subvert forcibly the cultural harmonization and the shrinkage of cultural repertoires. The "Odia word and thought of the month" project on Ornet is just the tip of the iceberg encompassing many larger processes of reclaiming lost cultural identity.

This is indeed one of the most important elements of ethnic online communities—this space to negotiate and document hybridized cultural identities forged out of the anxieties and conflicts of living a life between home and abroad. The instant connectivity of the Internet, the interactivity of email, and the archival capacity of database-enabled Web pages create a unique opportunity to document the shared narratives and resultant hybrid language, culture, and political constructs. As we explore the voices on Ornet, especially the discussion around the "Odia word and thought of the month" project, this aspect comes alive. There is a tendency to view the Internet as a space where Anglocentric information technology acts as a potentially hegemonic force that homogenizes the differences of culture and ethnicity of the users who feed into some master narratives of the content czars of cyberspace. Cyborg diasporas effectively challenge the possibility of such homogenization, and, interestingly enough, through the very power of information technology—interactivity, personalization, and hypertextual representation. The diasporas also challenge the deterritorializing tendencies of the forces of globalization. While pure virtual communities are born out of rebellion against constraints imposed by place, nationhood, and race, cyborg diasporas are created out of those very constructs of geography, nationhood, and kinship.

While cyborg diasporas effectively contest the cultural massification of the forces of so-called Western McCulture, they also resist the exclusionist cultural policies of their homeland. For many immigrants, a feeling of "neither here nor there" is the dominant way of being. Virtual networks are especially powerful when a small number of immigrants, such as the Odias, are geographically dispersed. They help these diasporas to have a critical mass of cultural, political, and economic influence. The work of the virtual communities as cultural units is not merely to contest the global master narrative. In fact, if one looks closely at Ornet, it is a project to differentiate the Odia identity from the other dominant regional cultures, such as those of the Tamils and Bengalis. Negotiating these multiple levels of cultural hierarchy enraptures the discussions on these virtual communities. Unfortunately, this often results in the fractures of the multiethnic societies from which these diasporas originate being replicated. The continued defensiveness of Ornet members to voices from outside the Odia-speaking Indian sphere is an example of a project of emancipation gone somewhat astray.

The role of cyborg diasporas in the contestation of global homogenization is certainly powerful, but it is not unique among the various localities that populate cyberspace. One of the reasons is the very design of the Internet to promote specialization and difference from the norm. Where cyborg diasporas do stand out is in the process through which they create community. They rise not *ab initio* from the bits and bytes of cyberspace and weak ties (Granovetter, 1973) of individualistic choice but from the very fabric of existing local neighborhoods, social kinships, and historical commonalities. The character of these existing neighborhoods is almost always that of non-Western urban communities, the spirit of which has been truthfully captured by Western communitarians, much less by sociologists of cyberspace. An angry younger member of Ornet tells an older member rather emotionally,

> I am forced to RETURN BACK YOUR LOVE AND BLESSINGS. I do not see any use of such love and blessings and I am sure I do not deserve to hang on to it any further. This does not mean that you cease to be my mausA [uncle] from today. Because that was your offering that I am returning back to you. I will still try to maintain and offer my respect as long as you are receptive of my offering. Everything else is immaterial to me.

He is not enacting readily codifiable transactions of "choice, voice, and exit" as Western rationalists would observe. Rather, he is enacting an age-old ritual of symbolic conflict between generations of uncles and nephews, between fathers and sons, and between older and younger brothers that are part of Odia culture.

These intimate relationships with the world outside and the incredibly complex ways in which it binds and unbinds the virtual community are the essence of the cyborg diaspora. While these "strong" ties give cyborg diasporas their color, they also infuse them with the venom of hypocrisy and conflict. This conflict, though a proxy for democratic engagement, is often very difficult to mediate through rational dialogue across computer screens. The double standards that are endemic to the angst-ridden existence of the online immigrants between the real and virtual spaces, between tradition and modernity, can be incredibly frustrating. This unpleasant spillover of the social malaise from Odisa society to Ornet cyberspace led one member to lament,

> We would prefer to forget Odia language and behave as "bideshi babu" [Mr. Foreigner], but not forget odia "kutaniti" [diplomacy], backdoor operation and dalari-giri, and "priya-priti-tosana" [nepotism] techniques. We would forget our dress code, but we would keep our "dalire au tike luna padithanta" [complaining] culture.

In the typical contradictory style of Ornet, the same day another member wrote,

> Orissa goes forward if we take it forward. Each of us has to dig through our own personality and feel the trait of sophistication that made Orissa, a land of bravery and culture. We must feel it through our hearts, minds and souls. We must dedicate a part of the daily routine to probe our roots and values. We must feel worthy of the soil that gave rise to structures of melody and harmony. We must feel the heart of the man with a beggar's bowl that carries a mind of purity and a soul of creativity. We must feel privileged that we have an opportunity to nourish it. We must feel humble to realize that we carry that heart.

There is indeed a great promise in the advances of networking technology and the intermingling of global cultures that can erode the constraints imposed on individuals by geography, identity, and politics. However, community life is also in danger of being overrun by the isolationism, exclusivity, and apathy that these forces are also capable of creating in us. Are we then stuck between these two extremes? Mulgan (1998), a leading British intellectual, ponders, "Are we doomed to a classical tragedy on which our love of freedom destroys our capacity to be interdependent?" (p. 3). We are yet to find ready-made answers to these difficult questions. Ornet struggles to bridge the divides that exist between the lived experiences in Odisa and in the United States. Many Ornetters reminisce about the warmth of interdependent relationships in India, yet they constantly question the impossible constraints that the Odia community imposes on individuals, notably the lack of tolerance for diversity in beliefs deviant from the norm and the "unfreedoms" characteristic of male-dominated gerontocracies. In the United States, the Odias celebrate the freedom to realize their intellectual ambition without restraint; simultaneously, they constantly challenge the rigidly held notions of unrestrained individualism where friendship is replaced with acquaintance and sharing is overwhelmed by reciprocal transactions. They continue to struggle with the extremes of social engagement, but through their debate, and through cyborg diasporas, there will emerge shared understanding and working prototypes of communities based on freedom, trust, and reciprocity. Such is the paradox and promise of community in an increasingly interconnected world.

Note

These insights into Odia culture and history benefited from Anustup Nayak's long discussions with Professor Bijoy Misra of Harvard University, a respected member of Ornet's virtual community and the "real" Odia community as well. Ornet's administrator and co-founder Ashutosh Dutta provided extensive information about the history of the community. This chapter also draws on the insights of Girija Kaimal and Prasanti Rao, both expatriate Indian women, who lent me the eye of psychology and politics.

CHAPTER ELEVEN

Working in Cybernetic Space: Diasporic Indian Call Center Workers in the Outsourced World

Ananda Mitra

THE PROCESS OF OUTSOURCING work from the developed countries of the North and West to the developing countries of the South and East is a phenomenon that has been happening for a long time. The process began as soon it was recognized that it was possible to gain access to cheap manufacturing processes and produce goods at a lower cost to increase profit. From the beginning of the twenty-first century, the matter has attracted increasing attention from media and politicians because contemporary outsourcing involves a process in which not only does the cheap worker produce a good to be sold in the West but also, the Eastern worker comes into direct contact with the Western consumer. This is precisely what happens when a person in Birmingham in the United Kingdom wants to make a railway booking and ends up speaking to a person in Bangalore, India. Using new technologies, call centers have made this process possible. However, this form of outsourcing opens up questions that are quite different from the ones that appeared when manufacturing jobs left the West. In particular, the outsourcing of service activities problematizes issues related to culture and intercultural interaction, where certain traditional vectors of power seem to be operating to create work conditions for call center employees that are akin to the diasporic condition of people from the East who migrated to the West. This chapter examines call center working conditions through the lens of diaspora.

The notion of diaspora has not necessarily been connected with outsourcing; the two constructs appear to be unrelated. Indeed, while outsourcing typically deals with the way in which tasks move from one place to another, diaspora deals with people moving from one place to another. The common factor between diaspora and outsourcing is the idea of place. Diasporic studies focus on the centrality of place to define how movement of people can lead to existential and ontological crises (Sheffer, 2003). In the same vein, discussions about outsourcing focus on the ways in which specific political and economic crises can be initiated as tasks and jobs move from one place to another. To examine outsourcing through the lens of diaspora, it is useful to conceptualize the notion of place and space, particularly in a world where those concepts are quickly being transformed by the adoption of computers in many facets of life. To be sure, many constructs, such as diaspora and outsourcing, can be recast in view of the ways in which the technologies of the Internet and the personal computer are reforming what it means to be in a place—real and virtual.

The New Spaces

At a most basic level, we become cognizant of location by constantly seeking an answer to the question "Where am I?" Our very existence can be threatened if we are unsure of our spatial location on some real map that points out the areas of peril and areas of safety (Shapiro, 1997). This desire is demonstrated in the development of advanced technologies as humans have constantly attempted to produce sophisticated maps to describe exactly what geographic space they occupy. This fascination for finding exactly where we are geographically located is now popularized even further with the manufacture and sale of global positioning systems (GPS), small instruments that tell us exactly where we are on the globe (Spaans, 2000). Maps also remind us where we are not only in terms of latitudes and longitudes but also in terms of the way in which we have chosen to carve up the globe according to defined political boundaries (King, 1996). To know where one is on the earth not only creates a sense of certainty with respect to location; it also reminds us that we are part of a large and complex system of classification where our location is tied to political, social, religious, and, ultimately, national categories, or nation-states (King, 1996; Wood, 1992). Fundamental to most political discourse is the recognition of the existence of the nation-state with well-defined and adequately protected geographic boundaries. In fact, the notions of diaspora and outsourcing become congruent because both focus on how people and jobs move from one nation to another. Space defined by geographic and political boundaries is thus a part of the popular cultural consciousness that eventually reminds us we are Americans or British or Nigerians precisely because we live our everyday life within certain boundaries and carry with us official identity documents that require us to live within those spaces and not leave them or enter a new space without adequate permissions (Shapiro, 1997).

These two notions of space are tied to an analogous mode of existence where, as Negroponte (1995) puts it, we live in the world of the "atom" and thus think in terms of objects, places, and spaces that are tangible and can be sensed. The entry of the computer, with its bit-based reality, however, calls the centrality of this system into question. It is now possible to create virtual spaces that can appear as authentic as the real spaces we have occupied. Thus we have witnessed the focus on "cyberspace" as an alternative to the "real" space we are familiar with.

The emerging virtual space has a set of characteristics that are important to consider. It is useful to conceptualize cyberspace as a discursive space where the key defining element are the texts and discourses distributed in an expanding computer network (Mitra, 1999). Indeed, little remains that is tangible about this space, and to experience cyberspace one needs to be immersed in the increasingly sophisticated discursive system new technologies are beginning to produce. This discursive space, even in its least sophisticated expression as Web pages, begins to appear very different from traditional, real spaces.[1] The increasing sophistication of cyberspace, primarily defined around how technologically advanced the discourses have become, has created a sense that it is removed and distinct from real space. What has emerged is a tension between the real and the virtual. Yet these two spaces are not truly distinct from each other. Indeed, I would argue that there is an interaction between cyberspace and real space that opens up the possibility of creating a new "cybernetic" space that did not exist before (Mitra & Schwartz, 2001).

Several different constructs coalesce when we think of the diaspora created by the new technologies that make far-flung call centers a viable commercial possibility. The people working in these places dwell in a cybernetic space precisely because their environment is shaped equally by the real and the virtual—the call center exists in a real space defined by a cubicle, through which the worker enters a virtual space defined by the discourse on the computer and the telephone. Furthermore, the real component of the cybernetic space is distinctly different from the virtual one, and thus there is a "virtual" spatial shifting not dissimilar to the traditional "real" movement that produced the diasporic condition. Thus it is useful to examine how the new diaspora and cybernetic space become connected.

Diaspora and Cybernetic Space

The diasporic condition is often the result of the way in which people are required to adapt to new spaces and the discourses that define those places. Traditionally, Diaspora has been used to describe the condition of the placeless people of Jewish faith being buffeted around the globe and always finding themselves "foreigners" in different lands attempting to produce a space for themselves (e.g., Tye, 2001). Not only were they encumbered with the need to find a geographic place for themselves; they also had to carve out an ideological space where the everyday practice of being Jewish could be maintained. The latter often is a process where the space is defined around discourses and there is a constant struggle to retain certain discursive practices that are quintessential to the diasporic people while also attempting to embrace the practices of the place of adoption. Indeed, this is the fundamental crisis of the diasporic existence, where frequently divergent practices need to be reconciled to maintain a hyphenated identity.

Consider, for example, how Jewish immigrants in and around Liverpool in the United Kingdom obtained a sense of safety around the perpetuation of the traditions of Jewish music and other cultural practices. Similarly, Gil and Vega (1996) documented the experience of Cuban and Nicaraguan families in the United States as they attempted acculturation in an unfamiliar place by balancing traditional practices with the ones that are demanded by the new place, and Lum (1991) considered the acculturation of Chinese immigrants in New York City as they attempted to find "familiarity" in and around the concentration of people of Chinese origin practicing Chinese traditions in the city. In all these cases the driving force is to retain certain practices and to adopt new practices.

In the end, the diasporic person who has been able to find a "comfort zone" and obtain a sense of the "good life" must be able to find a synthetic set of practices that help to negotiate the crisis of identity that diaspora necessarily produces. The physical displacement to a new place can make the process somewhat simpler because of the immersion in a new culture with its own discourses. For instance, in America, there are numerous examples of schools that have special classes for students who claim English to be their second language, and the institution attempts to help the diasporic to integrate into the new culture.

Many of these processes assume that diaspora is a process of movement in real space. The situation is very different when there is no movement in real space; a crisis of identity may occur where the new diasporic has to live in a new cybernetic space and thus to learn new discourses without being physically in the other real place. This somewhat paradoxical condition has been produced by the new form of virtual and telephonic outsourcing.

Outsourcing in Cybernetic Space

Much of the research on outsourcing has been conducted from an economic or business perspective, looking at the ways in which businesses have used outsourcing for financial gain. Some economists have argued that outsourcing refers to the way in which certain stages of a production process can be sent to a different country (Hummels, Rapoport & Yi, 1998); others have suggested that outsourcing is principally an activity performed by multinational companies that already have a global presence, and thus the idea of sending work elsewhere becomes more difficult to define. Grossman and Helpman (2002) have argued that outsourcing is more of a symbiotic process with the development of specific partnerships between corporations and countries so that both gain from the process. Most agree there are some fundamental aspects to outsourcing and that the principal descriptor of outsourcing is the movement of some component of work from a company's "home" location to a place "abroad." There could be many different reasons for shifting the location of work, but there is a strong spatial component to outsourcing. Indeed, this spatial component is the connection between outsourcing and diaspora: the diasporic condition deals with the movement of people; outsourcing deals with the movement of process.

From the American perspective, outsourcing is certainly not new. Indeed, from the moment it was recognized that there was cheaper labor in foreign countries and it was possible to send work abroad without too many regulatory or logistic challenges, not only did outsourcing begin but sophisticated economic models for understanding the process were postulated (e.g., Feenstra & Hanson, 1996).[2] The process of manufacturing remained acultural, and factories in countries such as Mexico and China were able to operate on their own terms provided the goods produced met the quality demands of the market in America. Indeed, the fact that foreign manufacturing units had few restrictions related to workplace conditions, pollution standards, and other rules that are strict and enforced in the United States opened the possibility of producing cheaper goods. The people who worked in the factories had little reason to understand or appreciate the culture where the goods would be used. Indeed, there has been significant discussion in the United States about working conditions in the factories in China that remain the primary provider of toys in the American market. To be sure, the child laborers in those factories are often oblivious of the conditions under which the toys made by them will be used by children in the West. That information remains irrelevant to the success of the process of outsourcing, since the key is the production of a good that will be transported and traded in a different real space from the space where it was made. In an atom-based economy that trades in real objects, the separation between the worker and the consumer was well defined and easy to police.

The situation becomes different as we move to a more bit-based culture and economy where the trading is no longer just in the realm of the manufactured good but in the regime of information and services, where the worker comes into contact with the consumer. Traditionally, such activities, often classified as the "service industry," operated under the assumption that it was important to have the service provider and the consumer close to each other spatially and culturally. For example, there was an expectation that when one needed to get one's telephone fixed, one would call a local phone number and someone in a local office would do something to get the telephone repaired.[3] There were often interesting consequences of this proximity; telephone operators and services learned about the people they served and there remained a cultural connection between the people simply because they were close together. The model of the telephone operator was adopted in many different service industries; the person at the other end of the phone would appear to be similar to the person seeking the service. Thus there was an expectation that the person helping with an airline reservation on the phone would have a familiarity with the destinations and might even sound like a neighbor. From the perspective of the service industry, working with information as the primary commodity, the options were limited to the space that was inexpensively connected by analog telephone technologies. From the perspective of productivity, it made sense for industries to recruit large numbers of skilled employees and place them in a central place where they would be able to field calls.

A change in this model began when corporations realized that there was an emerging space where the real and the virtual came together and the service could be provided much more inexpensively by a large number of people gathered in a central location that was no longer close to the physical space of the customers.

The service could be sent out, and the skilled people providing the service could be in a different place, often a different country. This option became attractive primarily because the new technologies made digital telephony boundaryless and affordable. The challenge was to find people who could be in cybernetic space on the phone: physically in one real place but culturally in another virtual place. This situation is no different from the diasporic, except that the physical location is now the place of origin and the virtual location is the place of adoption. In a bizarre way, outsourcing the call center turned diaspora on its head and created a new form of cybernetic diaspora where those working in call centers around the world had to adopt the culture of a different place when they entered the virtual world of the digital telephone and Web-based support.

Call Center in Cybernetic Space

The concept of a call center is not new. It is possible to trace the idea back to the 1970s, when, as discussed, American companies realized it was advantageous to centralize many telephone-based service activities in one location. No doubt changes in the telephone industry in the United States and the decreasing cost of long-distance calls made the idea of centralized call centers an economically viable plan. Broadly, two kinds of call centers emerged. There were those that were primarily responsible for answering incoming calls from customers. This was the model for industries such as airlines and catalog merchants, where trained people would answer calls and provide a service to the calling customer. In the second kind of call center, people would initiate a call to an existing or potential customer with the hope of making a sale. This process has also been called telemarketing. There are many differences in the way that these two forms of call centers are organized and staffed, but overall this distinction is still true for much of the call center industry.

When call centers were first established in developed countries, the presumption was that it was financially efficient to centralize telephone operations in a place that was physically in the developed country as well. The logic was to provide a branded face to geographically distributed customers through efficient service. The centralization allowed for better utilization of resources and greater quality control and standardization of the way calls were handled. However, with the development of better global telephonic infrastructure, it became clear that the idea of affordable "long distance" took on an open-ended potential. Technologies such as voice over Internet protocol, where a voice call is channeled digitally over the Internet, allow the costs of international calls to tumble and the clarity of calls to improve. The Internet thus made it conceivable for the centralized call center to move to a distant place without the customer even realizing that the call was being answered by someone 10,000 miles away as opposed to, perhaps, 2,000 miles away. As in many other cases, the Internet began to disassemble the traditional structures related to distance.

The fact that call centers could now be a-geographic posed a challenge of finding people who would be able to answer calls, particularly in the "in-call" call centers. There were several issues to consider: a reliable technological infrastructure; a cheaper skilled labor force than in the home country; and, for most American and British companies, employees culturally similar to the customer. In many ways, India provided a mix of conditions that made it an attractive option. The British colonial past ensured that a large portion of the people are able to communicate in English; India's post-Independence success in developing an efficient higher education system ensures high levels of skill among its educated population; and India's economic liberalization, beginning in the mid-1980s, led to the development of a strong information technology infrastructure. The two primary barriers for the American companies are the distance, with its consequent time difference, and the fact that the educated population in India tend to be more conversant with European than American culture. The latter issue is particularly important when American companies require the creation of a new diasporic population that physically lives in India but culturally is increasingly assimilated into American culture. What results is not the traditional "Americanization" of India but a more complex process whereby a particular segment of the Indian population experiences a set of ontological crises that resemble those often faced by people who experience the traditional diasporic condition through spatial movement. It is useful to consider the key issues surrounding the experience of call center employees in India.

Call Center Syndrome

Given that most call centers are operated for multinational corporations whose customers are located in the West, a set of interesting condition, results: the employees in the call centers have to invent a new set of practices to be able to work with customers in a completely different cultural space. Consider, for instance, the following statement by Gentleman (2005): "With employees working through the night to cater for clients in different time zones, the work requires staff 'to live as Indian by day and Westerner after sundown' and takes a 'heavy toll' on agents' physical and mental health." The idea that the people working in call centers are indeed living in two places is also visible in the ways call centers promote themselves. As a consequence of the need to align Indian lives with Western customers, the workers in India emulate the temporal rhythm of a different place. In a strange twist on the diasporic condition, the call center workers are now temporally placed in a foreign place while they still dwell in India. The odd work hours, arguably, are part of work culture in the industrial age, when factories never shut down, the day is arranged into three eight-hour work shifts, and the workers rotate through these shifts. In that scenario there is no expectation that all workers always work the "graveyard" shift. However, in a call center catering to Western customers, there is usually no other work shift than the late night ones. Consider the description of work hours provided by one who left the business: "I would return home at 2.30 am and go to sleep at 4 am. I would get up at noon and go back to work at 3.30 pm" (Mukherjee, 2004).

The better the worker is at stepping into the Western way of life, the more attractive he or she is to call center operators. Indeed, Gentleman (2005) makes this point in describing a call center employee:

> Vinita Rawat was a post-graduate student in English literature. She believes her fondness for Robert Browning and Jane Austen has given her an invaluable insight into British society, helping her penetrate the minds of the customers she speaks to 4,000 miles away.

The ability to understand customers is an important element of the successful operation of call centers, and not all workers have read Browning and Austen. Therefore, there is a need to "acculturate" the call center employee so that he or she has sufficient information to be able to empathize with the customer's cultural milieu. This is not very different from the way in which people who move from one geographic place to another learn the culture of the new place. In the case of the call center employee, however, the movement is a virtual one produced through the availability of new technologies that eventually require a set of tangible changes the employee must endure.

One change is taking on a new name when working at a call center. For the disaporic, this has been an important tool of survival from the moment of movement between countries. This is especially true with respect to Anglicizing names of non-English people. Accounts of name changes in the case of Jewish people and Africans in the Americas and Chinese people in Canada show that the diasporic go through a fundamental identity transformation as part of the acculturation process (Chen, 2004; Karpinski, 1996).

It is exactly the same process that the call center worker has to experience to be able to operate. Consider, for instance, the following account from the Web version of the popular American news program *60 Minutes*:

> New Delhi is nearly 11 hours ahead of New York, so manning the phones is largely night work. By day, the agents—as they're called—are dutiful Indian sons and daughters. By night, they take on phone names such as Sean, Nancy, Ricardo and Celine so they can sound like the girl or boy next door. "The real name is Tashar. And name I use is Terrance," says one representative.
>
> "My real name is Sangita. And my pseudo name is Julia," says another representative. "Julia Roberts happened to be my favorite actress, so I just picked out Julia." (CBS News, 2004)

What this illustrates is an ongoing and systematic process of identity transformation for call center employees. This transformation is essential to the operation of the call centers because the name makes a significant difference in the way the customer interacts with the service provider. The parallel between immigrants and call center employees in this respect is striking. Over the years in the United States, I have witnessed how a Sujata becomes a "Sue" and how children of immigrants are named with transnational names such as "Neal" that can be used across cultures.

For the Indian worker living in India, the name change results in a diasporic angst and other cultural transformations. For instance, Leslie (2004) reported the controversy over the outsourcing of work by the American state of Georgia:

> Tanneeru met with an employee at another call center that does work with private companies only.
>
> "What's your name?" Tanneeru asked.
>
> "My name is Khaja Kareemuddin," the employee responded. But like most operators in India, he said he changes his name and accent when he takes calls from the United States.
>
> "I'm a totally different person. I'm Derrick Andrews when I come to work," he said. "I try to put on a thick accent and talk so the people can understand me, and they should think that I am not someone from a third-world country."

According to state documents, the workers who take calls from Georgia food stamp recipients are college graduates who also change their names to make it easier for callers to understand them. This story points to another significant component of the acculturation process: language and accent.

In the diasporic condition, the issue of language often becomes particularly important. In some cases, it becomes imperative that the language of the place of adoption be learned quickly so that newcomers can "fit in." In some cases, the learning of a language is an institutional prerequisite for belonging to a place since countries like the United States and the United Kingdom encourage English proficiency as a precursor to obtaining citizenship. Furthermore, there is evidence to suggest that having the proper accent is as important in the acculturation process as having the appropriate language. To be able to speak like a local is an important achievement for those in diaspora, and there is anecdotal evidence that the children of the diasporic who are born and brought up in the country of adoption often are able to integrate better since they speak the appropriately accented language (e.g., Goto, Gee & Takeuchi, 2002).

This diasporic condition presumes a movement in real space, from one country to another, and an immersion in the new country, its environment, and its culture, making the acculturation process somewhat less daunting. However, for the call center employee, the acculturation must happen in the absence of immersion in or any real experience of the culture of the customer. Moreover, this virtual diasporic experience ceases as soon as the call center employee leaves the place of employment. Indeed, the call center employee experiences many of the anxieties associated with the experience of spatial diaspora, with the added twist of experiencing those anxieties in an artificial and virtual environment. Telephonic diasporic people must be acculturated through processes ranging from language training to creating a work environment and culture that attempt to mimic the culture of the customer. Consider, for instance, the way in which a leading international software company is dealing with the issue of acculturation:

> Indians working at Microsoft's new call center in Bangalore may sound a bit like they're from Seattle. For the past year, the center's 350 employees have been taught to speak more like Americans by Seattleite Andrea Koehler, a former University of Washington language instructor, who is part of Microsoft's training team. During a six-week language program, Koehler teaches the "technical-support professionals" to speak in a way that's clearer and easier to understand by U.S. customers who call for help with their Microsoft products. They also receive training in popular culture and current events to improve their conversational skills. Koehler likes the workers to listen to National Public Radio and watch Ken Burns's documentaries. But the workers prefer to get their U.S. culture from action movies such as "Independence Day" and sitcoms like "Friends." In the end they get a mix of things they'll enjoy watching and things that Koehler and Microsoft want them to learn. (Dudley, 2004)

Not only is there specific training in language and accent, but the new technologies have made it possible to offer such training in an efficient way where the call center employee is "immersed" in a discursive space and the Internet provides nearly all the cultural components, from American radio broadcasts to detailed local news stories. Call center employees are expected to live virtually in this discursive space so that they can be more like the people they are speaking to on the phone.

Sometimes companies offer a real environment that mimics the virtual. What is often described as an exciting work environment translates to tone that looks a little like the environment in which the customers operate. Consider, for instance, the following:

> It is not just money alone. The call centres offer an "exciting and pulsating" work environment with frills like cafeterias, gymnasiums, bowling alleys and excursions thrown in. Says Hena Alam of Keane Worldzen, "The culture of cafetarias and gymnasiums is a highly effective way of keeping the employees charged up. After the slog hours, it is party time." (Pathak, 2004)

Such environments often are attractive to the younger generation of workers employed at the typical Indian call center and may allow them to find a space where the "Other" culture and some of its attributes can be experienced; workers are better acculturated when they become virtually diasporic as soon as they enter a call center.

Despite their best efforts at language learning and accent changing, there persists a certain hostility, benign at some times and xenophobic at others, because real-life diasporics can never hide their Otherness, be it of color or of language. For call center employees, on the other hand, there is the advantage of being virtually diasporic; their racial and ethnic Otherness is virtually hidden. Thus all the training and acculturation of the call center employees is precisely geared toward making them not appear as the "Other" to the customer. Therein lies the potential of virtual diaspora, where the only contact is discursive—using the telephone or other tools, such as real-time chat sessions—where no "face" is associated with the call center employee. Thus, arguably, once the acculturation has been completed, there is a seamless cultural connection between the customer and the service provider.

The importance of this seamless connection is felt most when the connection is broken, at moments of crisis when the customer discovers that the person he or she is speaking to is not "Sue" but the owner of some unpronounceable name, sitting nearly 10,000 miles away. That is the moment when the diasporic Otherness of the call center employee is exposed, leading to the conditions that are harmful for the flow of international capital. It is at those moments that the customer is unwilling to work with that particular call center employee and perhaps the company that chooses to take its work abroad. Witness the conflict that arises when a call center employee accidentally shows the diasporic face:

> Partho let his accent slip and had to confess after being pointedly questioned that he was, in fact, an Indian sitting next to a telephone in Mumbai. The man told me, "You guys blew up the WTC," he said. "I tried to explain India had nothing to do with it, but he just banged the phone down." (Joseph, 2002)

Such outbursts demonstrate the large chasm between the two cultures that meet in this virtual telephonic space, and the results are not what the call center owners want. Often when callers recognize they are speaking to someone in India, they are quick to berate the person who is the "Other" as well as the institution that has forced the caller to experience this interaction with the "Other." The following story, reported by McPhate (2005), illustrates the reactions of an American caller:

> Debalina Das, 22, a computer help-line agent in the city of Hyderabad in south India, punched the button last winter for a call from the United States. The caller greeted her with a torrent of racial and sexual slurs, accused her of "roaming about naked without food and clothes" and asked, "What do you know about computers?" The diatribe ended with the comment: "This company is just saving money by outsourcing to Third World countries like yours."

This reaction is different from the way real diasporics have been treated, but it represents, I would argue, only a shift in the manner of response, and it potentially represents a phenomenon succinctly summarized by the manager of a multinational company in India: "Every time, it's racism only" (quoted in McPhate,

2005). In many ways, those who have been diasporic in real life would perhaps agree that the diasporic experience eventually problematizes one's identity, leading to an ongoing struggle to produce and retain an identity in balance between what one was and what one has become.

The Consequence of Virtual Diaspora

In the case of the telephonic and virtual diaspora, the identity negotiation is only temporary, since the diasporic experience is not all encompassing. Even though some call center employees claim that they enjoy interacting with people from another culture, this is an incomplete process because the interaction, and all the cultural baggage of the interaction, is interrupted the moment the person is off the phone and walks from the Westernized environment of the call center into the early morning streets of Bangalore. The traditional diasporic experience never ends so abruptly. Even people who try to maintain their cultural identity through the creation of mini ethnic communities cannot escape the environment of the new place as manifest in things such as climate, landscape, and general geographic characteristics. An Indian family that lives in Chicago may choose to remain truly Indian but cannot escape the snow outside. For the telephonic diasporic, there are no such markers in the environment that help them to fit in and do their job in a virtual place.

In some ways, this results in virtual diasporics having to shape themselves and their environments in different ways to manage the duality of their existence. This is a much greater challenge for the virtually diasporic because their lived experience remains very foreign from their workplace. The fact that their busiest period of work is at an unusual time of day poses some interesting challenges. For those who move from one country to another and experience the diasporic condition, the quality of life is related to the rhythm of life in the country they are in. Thus Indians living in the United States might well decorate their homes during Christmas and get accustomed to the American holidays and traditions. For the call center worker living in virtual diaspora, the experience is markedly different, as noted here: "It is 10.45pm, and in the dark streets outside preparations are being made for Diwali, the Hindu festival of light, but EXL works to English rhythms and even training sessions are conducted according to London time" (Gentleman, 2005). For call center workers, it is important to be aware of events such as the American Super Bowl or the weather in Leeds to do their job. In many ways, their diasporic condition becomes more challenging because their real surroundings never change but every night they are virtually somewhere else and struggling to fit in.

The issue of fitting in has been indirectly a topic of discussion in the popular press. The core issue that attracts significant interest is what happens to call center workers in the long run. Typically, working in call centers is cast in terms of a highly stressful job with no specific prospects of upward mobility for the employees:

> According to a study into conditions inside call centers conducted by a government-funded research institute, she is a prime example of an Indian "cyber coolie"—an expensively educated, highly intelligent graduate, who is wasting her talents performing exhausting, mindlessly repetitive tasks for the call centre industry, a sector which it claims offers no career prospects for the majority of its workers. (Gentleman, 2005)

Some researchers have considered the issue of fatigue from psychological and sociological perspectives, in terms of "work condition." For instance, Pradhan and Abraham (2005) argue that call center employees can suffer from multiple identity disorders and other psychiatric problems because they have to maintain two different identities. Such disorders can lead to long-term consequences, particularly if the person continues to work within the call center environment.

I would argue that many of the consequences of working in call centers can be perhaps better understood within the framework of diaspora. People in the diasporic condition often experience many of the same consequences as those working in call centers. The fatigue related to working in conditions that remain far more demanding than the convention of the known "old" world and the disorders resulting from having to live in two different cultures at the same time are considered a part of the diasporic experience. Call center workers are no different from many other diasporic workers in various parts of the globe. Therefore, those call center workers who feel they are closer to the other culture to which their virtual telephonic diaspora takes them often fare better in their work and perhaps are better able to deal with the stresses related to the "foreign" workplace. Consider the words of one employee: "I understand how things work in Britain; I know how the culture has developed. This helps me empathize with the people I'm talking to" (quoted in Gentleman, 2005). This is the voice of diaspora, where the outsider to the culture has an understanding of the culture he or she has to work in. Indeed, when the outsider is able to integrate into the culture, much like the traditional diasporic, there is a greater appreciation of the culture of adoption, as demonstrated by some call center employees. Describing the experience of one such employee, Desai (2005) writes:

> As Honey's example shows, the work of some must be exciting, stimulating or enjoyable. A few of them have considerable opportunities for self-development, and a handful may go on to compete with Americans in everything they do. It is a chance worth taking.

There is an internal contradiction in the experience of the call center employee. Much like the traditional diasporic, there is the interplay between living and working in a particular culture and trying to maintain the vestiges of another culture. Therefore, the circumstances of the call center employee can perhaps be better appreciated when looked at through the lens of diaspora.

Conclusion

In a peculiar way, looking at call center employees brings into question the nature of diaspora in the technological world. There are many important and striking similarities between the experiences of the traditionally disaporic and of those working in call centers. Typically, the diasporic condition has been considered a political, social, and cultural issue, and the scholarship explores the condition from the perspectives of race, oppression, Otherness, and similar concerns. In the case of call center employees, the research focuses on the issue from economic and psychiatric perspectives, because a key component of diaspora—movement from one place to another—is absent. I would argue that we now live in a technological environment where experiences that mimic the diasporic condition can be produced in the absence of physical movement. Assuming that a core component of the diasporic condition is the crisis related to culture, where the individual has to adopt a new culture either by completely relinquishing another or by striking a balance between the two, it is surely possible that such situations can arise in the new spaces we dwell in.

The need for a balance between the real and the virtual brings into focus the notion of the cybernetic world, where the real and the virtual synthesize to create a particular lived experience. Given the way that access to the virtual has changed, and because there is an increasing need and desire to live in the virtual, it could well be the case that, like call center employees, others too might experience diaspora in the virtual. The consequences might be similar to what the call center employees in India feel, and the solutions lie in discovering new coping strategies where the virtual and the real experiences are brought into greater alignment so that the cybernetic experience is harmonious and not internally contradictory. Perhaps this is what companies such as GlobalAMS are attempting to do in creating an Infocity in the state of Gujarat in India. On their Web site, the company describes the concept city as

> [s]elf contained with offices, residential, commercial and entertainment complexes. It would house an international school, hospital, banks, club, golf course, shopping malls, multiplex complex with bowling alley, disco/ktv, cyber café, squash and tennis courts, swimming pool, gym, go carting and much more. (Infocity, n.d.)

Here we see a vision in which the cybernetic experience will bring the West to India, whereas in a humorous urban legend circulating on the Internet the reverse is predicted to be true: in the future Americans begin to field calls for Indian customers:

> On this warm afternoon, the professor is teaching three ambitious students how to communicate with Indian customers.

> Professor: "Okay, Gary, Randy and Jane, first we need to give you Indian names.
>
> Gary, from now on, you'll be known to your customers as Gaurav. Randy, you'll be Ranjit. And Jane, you'll be Jagadamba." (Fun loving brat!, 2005)

The real Infocity and the fictitious professor in many ways represent the future of diaspora, which is just beginning to be seen in the lived experience of call center employees in India. As with many other things, technology is beginning to influence the diasporic experience itself.

Notes

1. Consider, for instance, the way in which it is now possible to "travel" to places by simply clicking on Web sites and seeing and often hearing what an exotic place is like. In the more sophisticated expression of this technology, as in the "Cave," a person can feel immersed in a new space through the technologies of virtual reality (Cruz-Neira, Sandin & DeFanti, 1993).
2. While the economic and wage debates continued and were increasingly politicized, the real cultural impact could be felt in many parts of the United States as entire communities disappeared when jobs moved to Mexico. Consider, for instance, the changes in the furniture industry in North Carolina. The furniture manufacturing industry was once one of the largest employers in the state. However, in the early part of the twenty-first century several factors, including the transfer of manufacturing processes to Mexico (and China), led to a loss of jobs (Nwagbara, Buehlmann & Schuler, 2002). This is but one of many examples of manufacturing industries seeing job losses as the work of making goods was moved out. In general, such moves produced benefits for the place the work went to by providing employment and the hope of a better economic future.
3. Telephone technology provides an interesting example of the local nature of this service industry. Before the advent of automatic switching systems, many telephonic connections were made manually by a telephone operator who physically pushed buttons and inserted wires to connect two people together. Spatial proximity was critical in this process; the technology was still at a stage where the operator, as a service provider, had to be close to the consumer.

CHAPTER TWELVE

Practices of Global Capital: Gaps, Cracks, and Ironies in Transnational Call Centers in India

Kiran Mirchandani

GLOBALIZED WORK processes have fundamentally altered the ways in which labor markets are organized around the world. While the globalization of work is far from a new trend, there has been much focus in the recent literature on the need to highlight how globalization is actually achieved. Critiquing the construction of globalization as an inevitable and irreversible process by which capitalism dominates nations, labor markets, and households, Bergeron (2001), for example, focuses on the "'gaps and margins' of the processes of global capitalism" (p. 999). Sassen (2001), in a similar vein, notes the need to "shift emphasis to the *practices* that constitute what we call economic globalization and global control" (p. 196). Freeman (2001) suggests that such an approach would allow for a rethinking of the hegemonic "masculinist grand theories of globalization that ignore gender as an analytic lens and local empirical studies of globalization in which gender takes center stage" (p. 1008). These theorists argue that grand theories that characterize globalization as a "meta-myth" (Bradley, 2000), a "rape script" (Bergeron, 2001), or a "narrative of eviction" (Sassen, 2001) do not sufficiently allow for an exploration of the incomplete and contested nature of the movements of capital and labor. Instead, the ways in which workforces are neither homogeneous nor passive with respect to globalized work relations needs to be highlighted. Such an approach emphasizes, as Sassen (2002) notes, the "cracks" that exist in the "wobbly political architecture" of social spaces.

This chapter documents the practices of globalization within a newly emerging transnational labor force—call center workers in New Delhi, India, who provide voice-to-voice service to clients dialing toll-free numbers in North America. In the past three years, India has installed reliable high capacity telephone lines in most of its major cities (*The New York Times*, March 21, 2001). As a result, according to a research report by the International Data Corporation, India is poised to register the highest growth rate in the call center services market in the Asia-Pacific region during 2000–2005 (*Asia Times*, June 9, 2001; Costa, 2003). In 2000 there were 500 foreign companies who outsourced work to about 60 call centers in India (*Economic Times*, June 19, 2001). By the end of 2003, India had 800 call centers (BBC News, December 11, 2003). Among the companies that have outsourced their call centers to India are British Airways, TechneCall, Dell Computers, America OnLine, GE Capital, Cap Gemini, Swiss Air, and American Express. Operators in these centers make telemarketing calls or provide service to customers calling about issues such as their insurance claims, credit cards, computer hardware, network connections, banking, and financial plans.[1] Calls are seamlessly and inexpensively routed via satellites across geographical spaces.

I begin with a review of literature on theoretical approaches to globalization that focus on the practices of transnational economic capitalism. Recently, theorists have noted that such analyses demonstrate that capitalism is continually under construction, that labor in the third world is heterogeneous, and that workers play important roles in relation to transnational corporate processes. Accordingly, I trace three practices that constitute transnational call center work—scripting, synchronicity, and location masking—and examine how these practices are under continual construction and negotiation by Indian workers.

Practices of Global Capital: Gaps, Cracks, and Ironies

Sassen (2001) notes that understandings of global processes have traditionally been limited to analyses of cross-border processes, such as international trade and investment. These analyses have produced a "rather empirically and theoretically 'thin' account" of the ways in which "the global economy needs to be implemented, reproduced, serviced, financed" (pp. 190, 192). To challenge the "transnational centrism" (Grewal & Kaplan, 1994) implicit in assumptions about the inherently powerful nature of multinational corporations and the inevitability of the spread of economic globalization, theorists have argued for the need to focus on the *practices* through which what is known as "globalization" is continually being constituted. These practices refer to the systems, norms, and work relations that structure workers' experiences and constitute "globalization." An exploration of the practices of globalization reveals the "intersecting effects and material consequences of so called globalization in a particular place" (Katz, 2001, p. 1214). As Sassen (2000) notes, "the global economy cannot be taken simply as given, whether what is given is a set of markets or a function of the power of multinational corporations" (p. 217). The shift of focus to the practices that constitute what we refer to as economic globalization allows for analysis of how diverse groups of workers play varied and active roles vis-à-vis transnational corporate and financial practices (Sassen, 2001).

The focus on the practices of globalization in this chapter facilitates a heightened awareness of two processes at play. First, it allows for an exploration of the active ways in which workers define and construct their work situations. As Gibson-Graham (1996) notes, "reading globalization...as a scripted series of steps and signals allows [us] to see the MNC [multinational corporation] attempting to place regions, workforces and governments in positions of passivity and victimization and being met by a range of responses—some of which play into the standard script and others that don't" (p. 132). Second, the focus on the practices of globalization highlights what Sassen (2001) refers to as the "cracks" and inconsistencies in global economic capitalism. Such cracks, which can be sites of the "hidden transcript" of power, may provide insight into the opportunities for political resistance (Scott, 1990). As Scott notes, "the social sites of the hidden transcript are those locations in which the unspoken, riposte, stifled anger, and bitten tongues created by relations of domination find a vehement, full-throated expression" (p. 120). Focusing on the gaps between the representation and experience of call center work, for example, reveals the incomplete and contested nature of transnational corporations in India. Appadurai (2000) challenges the view that globalization brings about a straightforward cultural homogenization. Rather, he argues, we live in a "world of flows" characterized by the constant movement of ideas, ideologies, people, goods, images, messages, and technologies (p. 5). Appadurai (1996) notes that "if *a* global cultural system is emerging, it is filled with ironies and resistances" (p. 29).

As Brah (2002) summarizes, "globalization does not exist in some rarefied stratosphere. It always touches the ground" (p. 26). Ethnographies of women and men working "on the ground" reveal the diversity of practices arising from globalized work processes. Freeman (2000), for example, provides a vivid illustration of workers' attempts to define their work and notes that "informatics workers in Barbados demonstrate through a variety of practices that they are not the passive pawns of multinational capital they have sometimes been depicted to be" (p. 36). She notes that women's jobs are a source of pride and pleasure and simultaneously a source of stress and dissatisfaction. She focuses on the agency women enact through their work and their lives. Freeman demonstrates that global capitalism is not monolithic; constructions of the "ideal third world worker" are both shifting and context specific. While other studies have revealed, for example, that young, childless, and unmarried women constitute ideal third world women workers, in Barbados family responsibilities are often believed to make women more committed to their jobs. Contrary to the assumption that multinationals seek a predefined flexible female labor force in third world countries, Freeman argues that ideal pools of flexible labor are actively and continuously created. In India, for example, multinational corporations are attracted not only by the large reservoir of cheap labor but also by the abundance of skilled and highly trained workers (Chhachhi, 1999; Nanda, 2000).

Rather than focusing on trade agreements, state policies, and corporate structures, this chapter highlights the practices of globalization as experienced by workers doing transnational call center work. The analysis explores "the intersecting effects and material consequences of so called globalization in a particular place to reveal a local that is constitutively global" (Katz, 2001, p. 1214). Following a brief discussion of the interviews I conducted and the nature of subcontracted call center work in India, I describe the relationships and processes that this work fosters.

Methods

The primary purpose of this project was to explore the nature of call center work within the context of global economic relations. Indian state policies encourage the subcontracting of IT-related jobs to India through tax relief, free trade zones, and incentives to private training colleges (Kaushalesh, 2001; Nayar, 2003). Types of IT service jobs now located in India include back office, remote support, medical transcription, call centers, database services, and content development (NASSCOM, 2001, 14).[2] While the call center sector is not the largest source of subcontracted IT service jobs, it is a sector that has seen considerable growth in the past three years. As Datta (2004) notes,

> The Indian Call Centre Industry has been growing at a mind-boggling growth rate of around 60 percent annually over the last 3 years. The revenues from this sector have grown from around $565 million in 1999–2000 to more than $3 billion by March 2004 and are projected to increase beyond $10 billion by 2006. Employment in this sector has increased from 140,000 in March 2003 to 200,000 plus by 2004. In fact it has been reported that the industry hired 200 persons every working day over the last one year.

In addition, unlike most of the other IT service jobs listed above, call center work involves direct interaction between Indian workers and North American customers. In this sense, call center work occupies a different place in the global assembly line than traditional manufacturing work. Taylor and Bain (2004) note that "the call center with its distinctive labor process can be offshored less easily than other non-customer facing, routine servicing activities" (p. 20). As noted in the popular press, "cultural distance is a bit harder to kill [when] company and customer are talking to each other on the telephone" (*The Economist*, April 29, 2000, p. 61).

Interviews were conducted in 2002 with two sets of individuals in New Delhi—call center workers and managers/trainers. All respondents were with organizations serving American clients. Call center workers were contacted via friends and colleagues in India and were employed by a variety of companies in the export processing zone (Noida) in New Delhi. All respondents were dealing with incoming calls (rather than telemarketing). Thirteen workers (seven men and six women) were interviewed. While most interviews were one to one, in some cases roommates or other call center workers were present in workers' homes where the interviews were conducted. Respondents were, on average, twenty-five years of age. One man was married, and one of the women was engaged to be married; all other respondents were single. All respondents had bachelor's degrees, and several had master's degrees or additional diplomas. None of the respondents had worked in call centers for more than one year (which is not surprising given the recent emergence of the industry). Workers earned between Rs. 5,500 and Rs. 10,000 (Can$150–$400) per month, with the exception of one male worker who had seven years of work experience and earned Rs. 30,000 (Can$1,200). A significant portion of salaries was made up of performance incentives.

In addition to call center workers, I also interviewed managers at three call centers, as well as representatives of three agencies that provide training for workers. These interviews with managers and trainers focused on the history of the industry, labor force demographics, and work processes. All interviews were tape-recorded and transcribed in full. Analysis focused on that which emerged out of the interviews.

Practices of Globalization in Transnational Call Center Work

Language and Scripted Taylorism: "Like a Keyed Toy"

One of the central mechanisms through which transnational corporations attempt to control the nature, timing, norms, and structure of work in India is through the use of standardized service scripts. As Leidner notes, workers engaged in interactive service work are often required to perform emotional labor as part of their jobs. Part of the work of providing service for a wage is the "management of feeling to create a publically observable facial and bodily display" (Leidner, 1999, 82).[3] Emotional labor is controlled by organizations through the development of detailed specifications for conduct (the scripting of "feeling rules") and the close monitoring of individuals' work. Call center workers perform emotional labor over the telephone. Given the synchronous ("live") nature of their interactions with customers, considerable resources are expended to develop processes within which the appropriate emotional labor can be facilitated, controlled, and monitored. As a result, the work practices in place at call centers in India can be likened to those in foreign-owned data-processing centers in the Caribbean, where "the open office is, at one and the same time, factory like in terms of its labor process and officelike in its muffled quiet ambience" (Freeman, 2001, p. 200).

Scripted service work in call centers is ensured through both monitoring and language training. Workers undergo both generic (accent, grammar, customer service) and process-specific (about the products) training before they are allowed to take calls. Some centers provide this training in-house and pay workers while they are being trained; others outsource the training to adult education sites, and individuals pay for the training they receive. Through training programs, transnational call centers engage in "language trafficking," which is the spread of particular types of English throughout the world (Swales, 1997). As Phillipson (2001) notes, English is a key instrument used by transnational corporations to break down national barriers. One worker describes the training program she was required to undergo:

> For the accent training we were being taught by cassettes. We had a special trainer—he was singing songs and listening to some conversations. And then we were made to see some movies and stuff. We were actually taught by cassette and we had to repeat all things like they do in nursery standards [schools], repeating the Aa, Puh, Tuh, Duh, and things like that. It was, you know, a bit funny at that time, we all used to laugh our guts out. What nonsense is this! You know at times you feel so frustrated. It's OK the way we speak is the way we speak. Why do we have to learn such stuff? Then we were told, the basic idea is that those people should understand you. (Female worker, respondent 13)

Many of the managers and trainers interviewed for this project objected to reports in the media that they were teaching workers to speak in American accents. As one manager notes:

> Voice training is not really developing an accent, it's neutralizing it...when you are looking at servicing an American client, you do not really need to sound American; however, you need to have a clarity of speech, and a pace of speech that is understandable by the other side. (Manager)

The justification provided for "neutralizing" accents draws heavily from discourses of human resource development wherein Indian labor is constructed as a flexible commodity that can be trained to meet client needs. As one trainer notes,

> Many Punjabi guys [sic] are having a Punjabi accent. Other guys are having other types of accents. So we don't need any kind of accent, we need a very neutral accent so that we can train them and get them the accent that our customers ask for. (Trainer)

Similarly,

> We have a phenomenally robust training system, by which even the least common denominator we convert them into a resource, that is, the raw material, we convert them into a resource which can face the customer in any part of the world. (Manager)

Workers, however, experience the training they receive as an Americanization of their English. The "neutral" in this sense contains a significant regional bias, reinforcing the "racist hierarchisation" implicit in identifying American English as legitimate and Indian English as illegitimate (Phillipson, 2001, p. 11). One worker notes:

> In India we speak English in a different, and in the States it is in a different way. [Interviewer: So they want you to learn...] A neutral accent. [Interviewer: What does that mean?] Neutral. Means they can understand what we tell. Like [for example] "schedule"—they say skedule.... And the American accent you have more r's rolling, there's a stress on the r's. So it's sem-eye-conductor, it's not se-me-conductor.... You're not supposed to speak anything except English, except American English. (Male worker, respondent 7)

> They have to tell us about the pronunciation part, about how to use an American accent. They actually make us listen to CDs, and they tell us constantly to watch CNN and all these channels. (Female Worker, respondent 11)

Nanda (2000) argues that while multinational corporations are often said to have strong alliances to their "home" countries, in fact, "MNCs' chief alliance is to profits, not to national boundaries" (p. 44). At the same time, the focus on developing "neutral" but American accents suggests that there is often a confluence between capitalist and nationalist goals.

Language training is accompanied by the emphasis on following predefined scripts. One worker who was undergoing training at the time of our interview noted:

> This is our script, we have to go through this. Thank you for choosing [name of American company]. My name is Tanya [assigned pseudonym]. May I have your first and last name. Thank You. May I call you by your first name? Thank you very much. How are you doing today?...These are the typical statements that we have to say—Great. Thank you. Excellent, Wonderful Job. These are the power words. We have to use those words in our scripts. (Female Worker, respondent 6)

While Leidner's (1999) research reveals that such service scripts can, in some situations, help workers to enforce their will over their customers and distance themselves from disagreeable interactions, the call center workers interviewed for the present project experienced scripts as deskilling, repetitive, and tedious:

> It's not that you are using your own words. You have to use these standard scripts. You have to use these same sentences...You're like a keyed toy...We were just told that we had to do the standard scripts. Just stick to your standard scripts. (Male workers, respondents 4 and 5 in conversation)

While workers are taught to emulate American culture, as Appadurai (1996) has noted, globalization seldom brings about a straightforward cultural homogenization. Bhabha (1994) argues, in his often cited essay "Of Mimicry and Man," that "colonial mimicry is the desire for a reformed, recognizable Other as a subject of difference that is almost the same, but not quite" (p. 86). Referring to the ways in which the English construct the Anglicized (that is, emphatically not English), Bhabha notes that "the discourse of mimicry contains an ambivalence; in order to be effective mimicry must continually produce its slippage, its excess, its difference" (p. 86). In this sense, mimicry, in producing a false copy of the original, makes a mockery of the colonial enterprise of engendering post-Enlightenment civility. Bhabha focuses on the ambivalence of colonial authority; similarly, while Indian workers are taught to mimic American work norms, there is a slippage between the information they are presented about Americans and the ways in which they interpret this information. In this sense, mimicry involves not only the colonizer's construction of the Other, but also the Other's construction of the colonizer. Specifically, through their discussions of their customers, workers construct Americans as rich but stupid.[4] In the context of the fact that workers are told that "customers are gods for us" (respondent 2) and "you have to be humble to the customer" (respondent 4), this rhetoric allows workers to pity rather than revere their American customers, thus subverting the ideology of the West as superior to the rest of the world (Hall, 1996).

One of the ways in which workers construct Americans as "stupid" is by making frequent reference to the lower value placed on higher education in the United States:

> Being a graduate in US can be a matter of prestige but being a graduate in India is just below average. Because average, average is even a post-graduation. (Male worker, respondent 3)

Workers note that nationality overrides class boundaries, which are being crossed with call center work whereby highly educated Indian workers employed in middle-class, white-collar occupations are serving often lower class, poorly educated American callers:

> Some Americans, they call [and] say, I want to talk to an American. Oh man, go on! You got an Indian and you are telling an Indian that you want to talk to an American!...Some of them, they really speak very very fast and that is a bit difficult...In any case we have to handle the calls. We can't say that, you are an American, we can't talk to you. Like they have the freedom to say anything but we can't say anything. (Female worker, respondent 2)

The notion of Americans as less educated than Indians is also reinforced through the training that workers are given. One worker says, for example, "Our CEO says that an average American is uneducated" (Male worker, respondent 3). Another worker describes scenarios provided during their training to illustrate responses workers should expect from American customers:

> They don't know anything about computers...If you say to them, just go to the start button, they will not be able to find the start button. Where is the start button?...And sometimes people are...talking about the trouble shooting steps and they're not sitting in front of their computers. [They say] I'm not able to see anything. And then we ask, Are you sitting in front of your computer? He said, No I'm not sitting in front of my computer. My God! One time [someone] called up and he said, you sent me a coffee mug tray and it was broken, send me another one...We asked our supervisor, was there any such scheme of sending in a coffee tray along with the thing?...Then [we realized] it was the CD drive! He used to put his coffee mug! We have so many examples like that. My God! (Female worker, respondent 6)

Many workers refer to the fact that while customers have little knowledge they have high disposable incomes:

> They don't know anything about computers. They put the [CD] upside down...We ask, OK how are you putting it in? The shiny portion should be down...Previously before we started interacting with Americans I basically had what I might refer to as [pause] it was in my mind, they are really good, they are really very intelligent, they have a lot of knowledge,

> nobody can beat the Americans. That was what my perception was. When I started handling calls the type of questions they ask, I said, Oh, it's bad. They only have money. They don't have brains. (Female worker, respondent 2)

Workers thus draw attention to the uneven development (which privileges national origin rather than education or intelligence) fostered by global capitalism (Wright, 2001). In this context, while they are often told that they need to speak in American accents so that Americans can understand them, workers interpret this as evidence of the parochial and erratic nature of Americans:

> The basic idea is that people should understand you...So that was the main motive behind learning all accent skills...Many a times people are very happy, and those people [say] how is it possible that staying in India[5] you can speak such good English?...But at times people are so rude—Oh, let me talk to someone who can speak English! I cannot understand you. We get customers like this also. One call, the customer is saying, oh you have fabulous English, you speak so well. And other call you get, oh my god let me talk to someone who can speak English. (Female worker, respondent 13)

Work processes and structures in Indian subcontracted call centers privilege the needs and sometimes racist perspectives of American customers. At the same time, workers' description of customers as stupid, uneducated, socially isolated, and erratic allow workers to pity or mock rather than revere Americans. In these ways, the focus on the practices of globalized work regimes provides vivid illustrations of workers' attempts to "live with industrial systems without losing [their] human dignity" (Ong, 1991, p. 296).

Synchronicity and the Myth of Globalized Time

Sassen (2000) notes that globalization involves the conflating of "national time and capitalist time" (p. 222). Indian call centers are at the forefront of such a confluence given the time sensitive nature of synchronous customer service work. As Adam (1998) notes, time is a "quantifiable resource that is open to manipulation, management and control, and subject to commodification, allocation, use and abuse" (p. 14). Indian call centers providing service to American customers are required to operate primarily during American daytime hours, and with the time difference of between twelve and sixteen hours between the United States and India, this means that call centers operate primarily in the night. Adam (2002) notes that such an arrangement signifies a "colonization with time" (p. 21) whereby Western clock time is exported across the globe and used as the standard. Indeed, India is promoted as an ideal location for the outsourcing of call center work from North America because the different time zones allow companies to operate in a seamless fashion around the clock. Adam (2002) notes that "in tune with the globalization of clock time, all that is local, context dependent, and seasonal becomes an obstacle to be overcome while particular histories and personal biographies are rendered irrelevant" (p. 17). Time is cited as a key factor in the literature advertising India's ideal location for IT outsourcing as it has "a virtual 12-hour time zone difference with USA and other major markets providing 24 X 7 business service platform" (NASSCOM, 2001). Time has economic value and is produced and exchanged. Such an approach, which constructs time as money, is based on the assumption that "capitalism has a built in clock that is always ticking" (Adam, 2002, p. 18). Typical shifts at Indian call centers are from 8 p.m. to 4 a.m., midnight to 8 a.m., and 4 a.m. to noon. Workers are sometimes picked up and dropped off in vans, which many workers greatly appreciate even though the vans operate on schedules that sometimes add between one and three hours to workers' time at the call center. Workers experiences of shift work, however, highlight the fundamentally "temporal features of living" (Adam, 1998), and the conflicts between local and global timescapes have a significant impact on their health, families, and unpaid activities:

> When I went [for this] job, I was very jubilant. I thought, OK, I'll study a lot, daytime is mine, so I'll be able to do anything. That's not possible. You're so tired. You're not able to do anything. Whatever time [you have] you sleep, even if you sleep for ten hours you don't get enough...you can never compensate a night's sleep...It's taxing, it's taxing, it's taxing on your social, it's taxing on your health, it's taxing on everything. (Male worker, respondent 1)

Call centers in India also operate seven days a week, through weekends and on Indian holidays. Workers have two days off per week, but these days only occasionally fall on weekends. Workers often mention feeling cut off from families and friends.

> I don't get time to talk to them [friends and family]. They all complain...If we work in the night shift, at least you get six and a half hours of sleep. You get up in the afternoon. You have three to four hours to do your work. But actually that is not possible. Because you get so damn tired after work. It's difficult to work in the night, that's what I'll say. (Female worker, respondent 2)

Worker shifts also change every few months, so they are constantly adjusting to new work times. While they are told that these constant shifts are necessary so that no one is permanently required to work the most difficult (all-night) shift, some workers mentioned that shift changes served only to ensure that they did not moonlight at other jobs. The constant shift changes have a significant impact on workers' health, as one worker notes:

> You need at least three to four hours of sleep in the night. That's what makes the big difference...That is what is making it difficult. I'm losing my appetite, I'm losing my weight...suddenly we were told we would be having our [shift] from twelve [midnight] to eight. It was very difficult to adjust in the first few weeks. Then I got adjusted to that time. Then again we were told that you're having your shifts from 7:30 till 4 a.m. And this shift, I find it very difficult to adjust. That is because I get home around 7 a.m., and it's very difficult to sleep in the morning because people wake up, they go around here and there. (Male worker, respondent 4).

Indian call center workers, in these ways, live and work in India but are expected to speak with American accents, take on American names, adopt American holidays and greetings, and work on American time. However, many workers identify the fact that they have to work at night as the most difficult part of their jobs, thus highlighting the fundamental immobility of time and their continued embeddedness within their local social contexts. While Harvey (1989) argues that one of the central features of global capitalism is "time space compression," whereby space is annihilated by time, the experiences of call center workers suggest an opposite trend. They experience what Katz (2001) has termed "time space expansion" in her ethnography of the effects of global restructuring in a village called Howa in Sudan: "from the vantage point of capital, the world may be shrinking, but, on the marooned grounds of places such as Howa, it appeared to be getting bigger every day" (p. 1224). Similarly, in transnational call centers providing synchronous service, workers are detached from the daytime-only spaces of social life such as markets, households, and transportation links.

Locational Masking and Situated Jobs

Part of the protocol workers in Indian call centers are required to follow is a masking of their geographical location. Workers are given American pseudonyms, and these are used as their names at work. They are also trained to avoid answering questions from customers about their location. Managers note that they often sign nondisclosure agreements with their American clients that require them to develop protocols through which their location in India is not revealed to customers. Such requirements exemplify the prevalence of "production fetishism" in the current economic order, which is, as Appadurai (1996) notes, "an illusion created by contemporary transnational production loci that mask translocal capital, transnational earning flows, global management and faraway workers" (p. 41). When asked why such locational masking is necessary and the purposes it serves, managers and trainers provide a number of responses:

> Why are we using this name [pseudonym]? Because it's easier to pronounce. That's customer service. (Manager)

> If we could achieve connectivity, nothing better than that. But if the guy sounds alien, you know, then my comfort levels are very low and I'm not going to impart any information. (Trainer)

> Most of the companies want to outsource their services to improve their productivity. There are various, I would say, concerns in [American] people [who may] say that they may lose their jobs. We want to minimize

> those effects. Because for sure when they're moving their work outside there are people who are losing their jobs there. (Manager).

> Some of the customers are wary about such things, because for example, there was a famous case of a multinational opening shop in India and they were not providing them with the right environment here. It was more of a sweatshop...customers may think [we are] a sweatshop, so that's one thing they want to avoid. It's bad publicity. (Manager)

These explanations and assumptions about American customers who are calling Indian call centers are used to justify the need to train workers in American accents and culture and to require them to use pseudonyms. Revealing that service work has been subcontracted to India may give rise to customer dissatisfaction for a wide variety of reasons, ranging from racism and ignorance, to concern about local jobs, to assumptions about exploitative transnational corporate practices. These concerns are not unfounded; workers report that they frequently face racism in their interactions with customers. The racism faced by immigrants to the United States is extended to Indians handling calls in India:

> One day a person went, you know, I don't want to speak to you. You have broken English. Please give me someone American. (Male worker, respondent 7)

> They call up, they say, I want to speak to an American. (Female worker, respondent 2)

> They may ask you, I bought a [product] and I should get support from an American, why should I get support from an Indian? (Male worker, respondent 5)

The experiences of call center workers suggest that Indians in India and Asian immigrants in America are named as equally "non-American." Locational masking, which, as the quotes above suggest, is only partially successful, serves to protect the interests of American corporations in light of the racism of their local customers. In fact, in attempting to reveal little about themselves, workers often reinforce negative stereotypes about Indians. One worker recounts the following incident in which a U.S. customer of Indian origin knew that he was likely talking to someone in India:

> [He asked] how is your relationship with Pakistan going? And things like, has Kashmir improved? You can't say anything. I told the...person, I'm not much involved in the politics. I don't read newspapers. (Male worker, respondent 4)

Workers are taught such strategies to minimize customer knowledge about the location of their work. In being asked to follow scripts and not reveal anything about themselves, however, workers are forced to reinforce notions of themselves as "keyed toys." At the same time, as Ong (1991) argues, worker resistances extend beyond direct confrontations with employers; "in manipulating, contesting, or rejecting claims [about their status] working women reassess and remake their identities and communities in important ways for social life" (p. 296). The dominant rhetoric adopted by government, trade organizations and media reports on call center work in India is that the industry provides highly desirable jobs. Workers evaluate these claims, and in doing so develop common understandings of the place of their work in the global order. The following argument about call center work was made in a 2001 report by the National Association of Software and Service Companies (NASSCOM) in India:

> Let's go back to the basics. A customer calls with a complaint or query specific to the product or service of the client company. The customer may get impressed with the speed or manner of response, but what he really wants is a satisfactory answer. That does not come from technology—it comes from knowing, not just the product, but the customer need, the market scenario, the real end benefit that the customer is looking for, and a familiarity with the marketplace...A call centre handling a tourism product must be manned by people familiar with the tourism industry, and in the same way, one handling process control instrumentation systems must be manned almost exclusively by qualified electronics engineers. (p. C28)

In this way, call center work is promoted as desirable and highly skilled. Transnational centers are housed in clean, well organized structures that often have entrances decorated with glass and marble. Van services for workers add to the prestige of the work. Workers employed to serve multinational corporations are paid up to twice the salary they would receive in local organizations. Accordingly, media and government outlets have identified call and back office centers as "India's new sunshine sector" (*India Today*, November 18, 2002). In particular, there is a notion call center work as a privileged occupation is created through the extensive screening process in place at the recruitment stage. Workers (all university graduates) describe being selected from hundreds of applicants and interviewed for hours before being offered the job. Workers reported:

> I was interviewed for six rounds with [the career consultant], then with [the call center] I interviewed for three rounds. Then I cleared the final interview, then I got the call. (Female worker, respondent 6)

> Around 200 people were shortlisted. And out of that 17 people were selected. (Female and male worker in conversation, respondents 11 and 12)

> [The interview took] seven or eight hours...one was the TOEFL test, and then they gave me a small objective type technical test, after that I was also given a one to one round, and then she gave me something to read out, maybe to see my accent, to see how I speak. Then I have a detailed questionnaire....I had again a one to one round with the technical people. Once I cleared that one, then I had a HR interview. After that HR interview in our company we get to be interviewed by a vice president or the CEO of the company...maybe a hundred people apply and only seven or eight or maximum ten are accepted. (Male worker, respondent 1)

Managers reinforce the rhetoric that call center jobs are highly skilled by referring to the higher status of call center work in India than in the West:

> One thing is there, in India, people take this job very seriously. I was abroad, so I know how seriously people take this job in the West [not very seriously, suggested by facial expression]. [Here] people take this job very, very seriously. They see a career path in this job, because it is a new industry, and in a new industry people move very fast. (Manager)

Despite having undergone a long process to obtain their jobs, call center workers are unanimously unconvinced by the arguments about the quality of call center jobs. Most of those interviewed for the present project noted that they do not anticipate remaining in the call center sector but had taken their jobs due to the lack of other job opportunities. Workers note,

> What happened with IT was that the balloon burst one day....Now if you just pick up yesterday's paper ... observe the four page ads, around 85 percent are for call centers....[People] tend to think it's a very glamorous job. In fact in my hometown [they say], OK you're working for a call center? Great! That's great! You're talking to American clients. But they actually don't know how tedious it is (Female worker, respondent 10)

> I know that call center is not going to last for long. It's very short term. I don't have any future plans with this call center. Not more than a year. Because there's no future. You can't sustain the taking calls throughout your life. It's just not possible. And this is no career. It's just a short term kind of job. (Female worker, respondent 13)

While they are paid high salaries in comparison with other service sector employees, workers note that call center salaries are far lower than professional ones:

> I have a buddy of mine, he is having four years of experience as a software developer and because of this slack he [had to leave]. He was earning around Rs. 25,000. And at this point in time [at the call center] he might be earning maximum Rs. 8,000. (Male and female worker in conversation, respondents 11 and 12)

Despite the fact that workers were repeatedly told they were fortunate to hold clean, white-collar, professional jobs with a multinational corporation, and that they are paid high salaries, the women and men interviewed for the present study repeatedly drew attention to the benefits that both American companies and Indian subcontractors extract from their labor. As one worker notes,

> If you think about the jobs there [in the United States], call center jobs, they would be paying at least ten or fifteen dollars per hour for a fresh person. So per hour means if you work eight hours you calculate, it would be around like a hundred and fifty dollars per day they have to

> pay. That's...like legislated minimum wage, not for people with any experience. (Male worker, respondent 4)

Many of the workers interviewed talked about the results of their mathematical calculation of the differences between their salaries and the amounts companies would need to pay if customer service was provided within the United States. A common set of figures was mentioned by various respondents, suggesting that this information was exchanged and discussed among workers. It was noted that the American company allocates $30 for each call handled in the United States, the Indian subcontractor receives $18 per call, and workers take an average of 20 calls a day. As one worker calculates:

> If you count yourself, within a single day you give your whole salary....So this is the call center industry—they're earning a lot. Exploiting, I can say, ninety five percent of the labor from the people...If you think about the jobs there [in the United States], call center [jobs], they would be paying at least ten or fifteen dollars per hour for a fresh person. So per hour means, if you work eight hours you can calculate, you know, it would be like hundred and fifty dollars per day that they have to pay. (Male worker, respondent 5).

Appadurai (2000) notes that we are living in a "world of flows" characterized by "objects in motion"; these objects include ideas, people, goods, images, and technologies (p. 5). Indeed, while call center workers in India are trained to take on American personas, learn about American society, and live on American time, they also gain an awareness of their connection to American labor markets and the global economic relations within which their jobs are situated. Workers situate their jobs in call centers within the wider political economy within which cost reduction rather than customer service drives the imperatives of capital accumulation and profit maximization (Taylor & Bain, 2004). Such awareness allows workers to challenge employer definitions of their work as privileged, skilled, and desirable. Instead, they note that even experienced, skilled workers earn far less than their American counterparts, thus naming geographical location as the prime determinant of the value of labor in the global economy. Rather than becoming the "ideal Indian workers" promoted in the state literature in the face of the constant threat of capital flight, workers see their work in call centers as transient and are in constant search of opportunities for more fulfilling work.

Conclusion

This chapter highlights transnational corporate practices in globalized call centers. These practices are continually under construction, suggesting that global capitalism is neither inevitable nor predictable. In focusing on the practices of globalization, this chapter contributes to the work of a number of theorists who have stressed the importance of situating analysis of the "gaps" and "cracks" within broader understandings of the political economy of globalization. As Grewal and Kaplan (1994) argue, it is important to move away from the "transnational centrism" in analyses of global capitalism that assume transnational corporations have invincible organizational structures and possess inevitable hegemonic control. As Gibson-Graham (1996) notes,

> if we create a hegemonic globalization script with the MNC, the financial sector, the market, and commodification all set up in relations of mutual reinforcement, and we then proclaim this formation as a "reality," we invite particular outcomes....By querying globalization...we may open up the space for many different scripts. (p. 145)

For example, in drawing parallels between their jobs and call center jobs in North America, and between service and professional work, Indian workers defy their construction as a passive and grateful workforce, thus "crafting" themselves in light of the "shifting fields of power" (Kondo, 1990, 260).

There has been considerable media coverage of transnational call centers in India. Many reports celebrate the new growth in subcontracting with euphoric enthusiasm and claim, as in an article in *The Statesman,* that it proves that the "Age of India Cometh."(March 24, 2004) The author of this article notes that

> Money for wages will flow into India in significant amounts, much of it directed into the pockets of India's growing middle class. A whole stratum of the population will earn salaries which, by US or British stan-

> dards, are low and therefore competitive, but which by Indian standards are, if not princely, more than comfortable....This inflow of money into wages has positive implications for the Indian economy as a whole. (p. 5)

In other reports, such as one based on the views of Praful Bidwai, who argues that transnational call centers reduce young, highly educated Indian workers to "cyber-coolies," the subcontracting trend is characterized as "a perfect sweatshop scenario, except that you're working with computers and electronic equipment rather than looms" (BBC News, November 11, 2003).

The analysis in this chapter suggests that these dichotomous perspectives largely fail to capture the experiences of Indian call center workers. Neither are workers grateful for and satisfied with the so called comfortable jobs that allow them to escape the perils of joblessness, nor do they want the subcontracting trend to end, despite their recognition of the many negative professional, social, and physical effects of their jobs. Rather, for call center workers, as Sarker and Niyogi De (2002) argue, "resistance and social change arise...from an entanglement with regimes of dominant knowledge/power, not outside them" (p. 2). Workers object strongly to the routinized and deskilled nature of scripted service work, and their resistance is largely directed toward attempting to improve the quality of their jobs.

There are signs suggesting that worker demands could begin to shift the practices of globalization that currently characterize transnational call center work. Following the partial closure of call centers in India by three large transnational companies (Capital One credit card services, Dell computer support, and Lehman Brothers investment services), allegedly due to the poor American accent skills of workers, for example, there was a lively discussion in the Indian press of the implications of the cancelled contracts. Interestingly, rather than discussions of the need for better training to prevent further loss of contracts, *The Times of India*, India's mainstream national newspaper, invited readers' comments on the question "Should Indian call center workers retain their accent and identity?" An article in *The Economic Times* provided the following response:

> So, should [Indian call center managers] start buying Hollywood DVDs by the armful and ensure that everyone on the premises, from the security guy to the CEO, starts talking the walk with an American accent (never mind that there is no such thing as a single American accent)? No. Non. Nada....Here's why:
>
> —There is so much diversity in the US anyway that few raise an eyebrow at hearing a non-American name or accent in their everyday exchanges. Check in at a motel and you're virtually guaranteed to hear Gujarati-laced American.
>
> —There's no way that the average American call centre operator is better equipped than his or her Indian counterpart to answer queries and provide support over the phone, especially on technical issues. India's call centre operatives are mostly college grads, which is not the case in the US.
>
> —India's call centre operations have global clients, not just American clients. One accent won't serve all. In any case, how long can customers around the world be fooled into believing that their calls are being answered by someone in their own country? And why is that so important anyway? (July 13, 2004)

Highlighting the continually contested and heterogeneous nature of global capitalism would allow for further analysis of the microprocesses within which resistance to transnational subcontracted work occurs as the industry develops in India. As Taylor and Bain (2004) note, "although the Indian call centre industry is relatively new and research is still in its infancy, evidence is emerging that, just as in the west, the call centre is contested terrain" (p. 20).

Notes

This project was funded by the Shastri Indo-Canadian Institute and the Social Sciences and Humanities Research Council of Canada (grant number 451915). I would like to thank the interview participants for their enthusiastic and generous involvement with the project. An earlier version of this chapter was presented as a paper at the 2003 Critical Management Studies Conference. Also previously published as Kiran Mirchandani, "Practices of Global Capital: Gaps, Cracks and Ironies in Transnational Call Centres in India," Global Networks, Blackwell Publishing.

1. There has been considerable research on call center jobs in a variety of countries. Examples include Bain and Taylor (2002) (comparing British, Dutch, and U.S. call centers); Mitter (2000) (Malaysia); Barnes (2004) (Australia); and Larner (2002) (Canada and New Zealand).
2. Call centers are either inbound (customer service) or outbound (telemarketing/sales). A significant percentage of call centers in India are out-

bound. As discussed in the Methods section, the focus of this chapter is on the inbound centers providing customer service.

3. Although Leidner does not refer to phone-based interactive service work, voice services, too, can be assumed to involve the management of feeling.
4. A few workers noted positive perceptions of Americans (e.g., that they were more patient, more willing to solve issues via the telephone), but these were by far the exception.
5. As mentioned earlier, locational masking is not always successful, and customers sometimes guess the workers' location.

CHAPTER THIRTEEN

"Around Sourcing": Peripheral Centers in the Global Office

Divya C. McMillin

THE WORD "outsourcing" is an emotionally charged one, with the potential to trigger waves of alarm among conservatives and middle-class workers in core countries. It drums up paradigmatic images of closed factories in Middle America and bustling call centers in India and China, of tense despair on the faces of white Americans and happy grins on those of third world employees as they achieve the American dream so desperately sought after by their counterparts in the United States. In the world of technology and information services, India in particular is deliciously perched on the periphery, the blushing bride offering up millions of university-educated, English-speaking graduates eager to earn more in the private, transnational information industry than they would in government-backed or even private local industries. So tantalizing is this venue for its seemingly limitless supply of willing technoworkers that *Newsweek* devoted a March 2006 issue to "the new India," represented, tellingly, by a blouseless sari-draped dusky long-haired woman with hands folded in the classic *namaste* greeting. The analogy of the giving female nation is apparent.

Overshadowed only by the current, deeply divided debates and demonstrations on illegal immigration in the United States, where the aliens have indeed landed on home soil and taken on jobs Americans won't do, outsourcing continues to be an object of anxiety, where aliens devour jobs Americans *will* do, without ever leaving their countries. And now even North American college and high school students are encased in a rhetoric of vulnerability where their homework help can come from university graduates in Indian cities such as Cochin or New Delhi rather than the exponentially more expensive tutoring services within the United States (Paley, 2006).

As we may suspect, the technoscape of information outsourcing is far more complex than a simple subversion of power differentials where peripheral citizens are secure in jobs that have displaced a core proletariat. Neither is it the win-win global office that emerges from Friedman's (2006) observations in India, where computing companies such as Satyam Online "aroundsource" to villages, "breathing economic life" (p. 3) into them by having villagers support services required by Indian cities, which in turn support multinational corporations overseas. Immediate economic benefits notwithstanding, the small-fish-feeding-bigger-fish structure provides a hint of the inequities built into the system.

Drawing from ethnographic fieldwork among employees in multinational call centers in Bangalore city, this chapter assesses the neocolonial processes that transfigure the average Indian worker into the ideal service provider in the global marketplace. As the relatively young (eighteen to forty-five years), urban and semiurban call center proletariat pours its education, skills, time, and energy into the service of a remote bourgeoisie for comparatively lucrative local pay yet globally substandard work conditions, certain pertinent questions arise: How does the cultural transformation of this labor force take place? What local structures support this transformation? And what implications will this process have in the long term?

To answer these questions, participant observation was conducted during summer 2004 in two call centers, and in-depth interviews were conducted with forty employees from six call centers in Bangalore. The respondent pool was deliberately kept small to allow for in-depth interviews and the collection of other qualitative data. Newspaper, radio, and television reporters associated with call center stories were interviewed, and extensive information was derived from media and public libraries in Bangalore. Call centers in this city are housed in multipurpose buildings such as the gleaming International Tech Park (ITPL),[1] complete with its granite-façade waterfall, gym, coffee shop, restaurants, and Thomas Cook and American Express currency exchange offices; in parent businesses such as Hewlett Packard (HP), Dell, or HSBC; or in dedicated venues such as the MSourcE complex, which also houses restaurants and a coffee shop for its employees. Tight security prevents casual visitors, and armed guards cluster at entrances, interrogating anyone without an entry permit. While participant observation was conducted at the ITPL, interviews were conducted off campus because most employees were hesitant to converse about their jobs at their place of work. Access to the ITPL was easier because the currency exchange offices and restaurants are open to the public, but access to other venues was gained through inside contacts, with a visitor's permit that strictly enforced a limited time for participant observation and interviews[1].

The use of words such as "core" and "periphery" in the introduction reminds us of the good old days of elegant world systems theory when the neat ordering of the world according to economic differentials and situational dependencies dictated who would have and who would not. Rather than dismiss this model as many theorists of globalization tend to do for its failure to address the interactions and interdependencies that define particularly technoflows in a networked world, we revive here Wallerstein's (1990a) evaluation of the agency of the individual in a system entrenched in structure, no matter how dynamic and complex the flows between core and peripheral communities. He writes (1990b) that "in virtually any social situation, the actors may be ranked in a hierarchy of power—some stronger, some weaker—it follows logically that the stronger 'get their way' more frequently than the weaker" (p. 65). The notion that outsourcing somehow leads to autonomy and agency in the third world worker and that we have finally arrived at a brave new flat world (Friedman, 2005) is a misguided one in the face of a variety of examples—the call center being one—of the enduring structures of economic and political inequity.

Call Centers as Hybrid Spaces

After economic liberalization in 1991, the urban Indian landscape exploded with foreign and private investment in industry. In the television industry, the early 1990s witnessed the phenomenal rise of private foreign and regional television networks, providing stiff competition to Doordarshan metro stations in urban areas (McMillin, 2001). In the mid-1990s, mobile phone connections rapidly replaced landlines, allowing urban Indians to access the Internet, completely bypassing the cumbersome and inefficient methods offered by the Department of Trade (Miller, 2001). In 2000, dot-coms appeared at the average rate of three per day, and by the end of the year, there were an estimated 50,000 India-specific Web sites. Many of these crashed soon after because they lacked viable customer-oriented business plans (Ahmad, 2001). Following the short-lived dot-com boom in India (2000–2002), IT-Enabled Services Business Process Outsourcing (ITES-BPO) was first considered by many infotech industry watchers as a capricious venture, liable to crash because of poor infrastructure. By 2001–2002, in an environment where technological innovations raced ahead of government policy and regulation, call centers in the country emerged as a lucrative site for transnational transactions. The success of BPOs, particularly call centers, continues at the time of writing. By 2008, the Indian ITES-BPO market is expected to reach $20 billion, providing jobs for 1.1 million (NASSCOM, 2003).

Far from being homogenized by global cultural flows, as has been posited by theorists who support technological determinism (Schiller, 1991), third world cities are witnessing the creolization of local culture (Foster 1991). Such creolization or hybridity refers to the liminal space between global and local where, as Bhabha (1994) writes, this "interstitial passage between fixed identifications opens up the possibility of a cultural hybridity that entertains difference without an assumed or imposed hierarchy" (p. 4). However, as discussed elsewhere in the context of television in India (McMillin, 2003), the hybrid is certainly hierarchical; call centers stand as strong symbols of a neocolonialist environment, where laborers need to enter into the cultural contexts of their employers and clientele based in the United States, the United Kingdom, Germany, or the Netherlands, as the case may be, and, using their knowledge of the range of customer services available to the client, converse fluently, stripping away as far as possible any indicators of their local Indian contexts. It is true that local indicators such as an Indian accent or name are more acceptable in contemporary call center interactions—but only insofar as they do not interfere with the provision of service. Just as call centers in the United States are located in rural areas because of the high cost of real estate in highly populated areas (Sharp, 2003), on a global scale developing countries offer low real estate, labor, and infrastructure costs, making them prime targets for call center location. What differentiates a neocolonialist environment from a colonialist one is the context of globalization, where the focus is not on overt force and imposition but on interconnectivity.

The Call Center as a Global Office

ITES involves the outsourcing of processes that can be enabled through information technology. This means ownership and management of the process is transferred from the customer to the service provider (NASSCOM, 2003). Call centers are synonyms with ITES and are defined as departments that respond to all sorts of inbound and outbound business-to-customer (B2C) and business-to-business (B2B) communications. They have existed in some form or other since the evolution of the modern business structure. However, it is only within the past decade that call centers have become a primary way through which companies conduct their business, particularly because of the significant role played by communication technologies (Sharp, 2003). Business process outsourcing is defined simply as the movement of business processes from inside the organization to external service providers (Click & Duening, 2005). Examples of services provided by call centers are customer care, Web sales and marketing, billing services, database marketing, accounting, transaction document management, transcription, telesales and telemarketing, benefits administration, tax processing, HR hiring and administration, and biotech research (NASSCOM, 2003).

Transforming the Peripheral Subject

For call center employees to live and breathe the cultural contexts of their clientele, names are changed from Indian to Western ones (particularly for U.S. clients) and fictional personal profiles are developed with residential roots in some prominent city in the United States. Of the forty call center employees interviewed, 50 percent were male and 50 percent female. Sixty-six percent (twenty-five) were customer service representatives (CSRs), of whom 88 percent (twenty-two) were between twenty-one and twenty-five years of age. This reflects Singh and Pandey's (2005) study of 100 call center women in Delhi, of whom 67 percent were between twenty and twenty-five years of age. Of the CSRs, 56 percent made between Rs. 5,000 and Rs. 10,000[2] ($111–$222) per month and 20 percent made between Rs. 11,000 and Rs. 15,000 ($244–$333) per month.

CSRs form the bottom rung of the call center structure and can be promoted to senior CSR on the basis of their performance ratings. Team leaders form the next rung of the hierarchy, and, drawing a salary between Rs. 15,000 and Rs. 35,000 ($333–$777), supervise teams of around fifteen CSRs. Process coaches, who supervise around teams teams, also draw salaries between Rs. 15,000 and Rs. 35,000 and are usually compensated at the upper end of this scale. An operations manager oversees process coaches and teams and receives a salary of around Rs. 50,000 and Rs. 100,000 ($1,111–$2,222) per month. The operations manager reports directly to the vice or assistant president, who is paid between Rs. 100,000 and Rs. 300,000 ($2,222–$6,666) per month. The president of the call center can make as much as Rs. 500,000 ($11,111) per month.

Most of the CSRs in this analysis said they liked their jobs because of the interaction with people, challenges in sorting out client problems, young colleagues, comparatively good salary, free transport, and the opportunity to improve communication skills and English language and diction. Many stressed the playful work atmosphere and said the modern "MNC looks" of the workplace made them feel they were stepping into a global environment, a space that was quite different from their homes. They particularly enjoyed talking to Americans and other foreigners across the world, learning the American accent and culture, and breaking cultural barriers through their interactions.

Ashok, a twenty-four-year-old CSR whose call name was "Brad," said the frequent parties at the workplace made the environment fun. He said:

> Each team has a name and there are lots of parties, lots of gatherings. And we do the Indian and the American holidays. On July Fourth we decorate according to the American flag. We celebrate that Americans achieved their independence and we learn about George Bush, and we just celebrate—we wear red white and blue.

Ashok considered his name change to Brad as a necessary requirement of the job to get to know clients and their culture. Although speaking in an American accent was part of his work, he could switch to his Indian accent when work was over, and this did not interfere with his sense of Indianness. On the other hand, Harish, a twenty-nine-year-old assistant manager of operations, said:

> Our cultural identity is definitely affected. Although we do have "India" days, there is a large influx of Western and U.S. culture. That is not necessarily bad. But it depends on the individual. Some call center people continue this accent and mannerisms—I've seen a lot of people who think and act American on the outside as well and anybody can tell, "Oh, this person is working at a call center." On the other hand, you can learn a lot of things; you learn to be polite on the phone—it is actually a delightful experience. Pseudonames are going away probably because a lot of customers in the U.S. and U.K. know the calls are going to India. Some crib; some are happy speaking to an Indian and are impressed with the technology.

The respondents mentioned that in addition to their high salaries, they also could earn more through extra work. Singh and Pandey (2005) detail the remuneration schedule of the workplace. With a five-day workweek, employees are allowed two to three days of casual leave per month that can be accumulated to a total of eighteen days over six months. In place of casual leave, the employee may choose extra points to be added to her or his performance to receive further incentives. Work on public holidays may result in more points, so that in a month the employee may earn up to Rs. 3,000 ($67) extra.

The selection process for CSRs typically includes phone and email screening, behavioral interviews, testing for keyboarding and written communication skills, evaluation of sales and customer service aptitude, and screening of references. Mohan, a thirty-year-old back office senior executive, stated that applicants are short-listed based on experience and expertise. Selected employees then receive training for one-and-a-half to two months on the product and software, after which they "'Go Live' and are on ninety days' probation. Depending on the impression you make [and] your quality scores, you get the project for two to three years."

CSR training typically includes workshops, seminars, call observation, product knowledge tests, monitoring and coaching, online tutorials and toolkits (Sharp, 2003). Santosh, a twenty-seven-year-old processing executive, discussed accent coaching as an integral part of the Indian call center training process. Recruiters specifically sought those who were exposed to U.S. culture either through television or their urban residence. Those with strong native accents were eliminated. Women proved better at picking up accents than men and were preferred in CSR positions. As with Singh and Pandey's (2005) study, this analysis showed that there was no discernible pattern of gender discrimination at the CSR level. Women held most of the mid-level management positions. Although no higher management executives were interviewed, the respondents stated that men occupied the senior positions in their organizations. Of the CSRs in this study, at least 79 percent (thirty) had bachelor's degrees in either the arts and sciences (47 percent) or commerce (32 percent).

The non-CSR respondents (34 percent) occupied various levels of middle management (assistant manager, two; senior customer executive, two; unit manager, two; technical quality assistant, one; operations manager, one; customer care consultant, one; team manager, three; processing executive, one; CSR training, two). These executives had achieved their positions either by working their way up over the course of three years or more or by virtue of their qualifications as engineers or MBAs.

Shalini, a twenty-seven-year-old team manager, explained that her job was not just to meet call response targets but also to enforce a clothing policy that required only closed footwear, no skirts above the knee, and generally formal wear on weekdays, with casual wear on Fridays and Saturdays. Team managers also had to plan and conduct orientations and team activities such as potlucks at the homes of the vice presidents or operations managers. Teams often went shopping, to the pubs, or on picnics together. For many young adults, particularly those who do not have families in the city, the call center, said Shalini, "becomes a second support system. We stay on even after work and go out for *iddlis* for breakfast in the morning after our night shifts." She said call center applicants and employees were encouraged to watch American programs such as MTV countdowns, *Friends*, *King of Queens,* and other sitcoms to familiarize themselves with family contexts and American accents. Interestingly, of the respondents who discussed their television watching (almost 50 percent said they had no time to watch television), most listed American programs such as *Hot 'n Wild*, *Friends*, *Oprah*, and *CSI* and foreign channels such as Discovery, National Geographic, Animal Planet, and Cartoon Network as their favorites.

With high turnover rates (between 22 and 50 percent globally), call centers allocate as much as 60–70 percent of their budgets to staffing, and around 72 percent of companies use external recruitment agencies to keep up staffing (Sharp, 2003). Retention strategies include health awareness workshops for employees, calorie charts to encourage healthy diets, exercise routines, including yoga, mentoring and leadership programs, vehicle loans, medical insurance, provident funds, and various investment opportunities. Yet the attrition continues to be high, providing a clue to the challenges of call center work.

The Center Cannot Hold

What makes the call center an intriguing site for the study of transnational processes and identities, in addition to the transformations it requires in terms of accent and persona, is its nocturnal function. The fact that India is between nine-and-a-half and thirteen-and-a-half hours ahead of the United States, between four-and-a-half and five-and-a-half hours ahead of the United Kingdom, and around four-and-a-half to nine-and-a-half hours behind Southeast Asia and Australia necessitates call center operations during the night. For middle- and lower-middle-class conservative families to give up their young daughters, wives, sons, and husbands to a night trade can only mean that such a trade is extremely lucrative. As Verma and Sharma (2003) explain from their study of 100 adolescents and their families in urban India, urban and rural girls learn, early on, that the public sphere is for the most part a hostile, male-dominated space. Cultural transformation of call center employees, then, occurs not just in individual appropriations of Western accents, clothes, and interpersonal behavior but also in family structures and norms. Young females claiming the night as a time of work are no longer derogated but glorified.

Yet the respondents in this study reported working during the night while their family routines continued during the day as the most significant drawback of the job. A majority (53 percent) worked a nine-hour shift, while 29 percent worked an eight-hour shift. Some (8 and 3 percent) reported working ten- and twelve-hour shifts, respectively. Of those who responded to a question regarding the breaks they received (80 percent), most (91 percent) said they received a total of one hour's break. This was distributed over a thirty-minute lunch break and two fifteen-minute coffee breaks.

Respondents were candid in their discussion of what they disliked about their jobs. Alphonse, a twenty-four-year-old technical quality assistant, said he was not able to be home with his family during important holidays. Anil, a twenty-year-old CSR, said, "When the call flow is high, group leaders do not let us take breaks. They do not ever let us keep the line busy for a while even to make notes on the calls." Anita, a twenty-five-year-old CSR, said she would quit very soon because the night shift was wearing her down. Goutham, a thirty-year-old assistant manager, said the shifts were particularly brutal for married people with young children. The stringent log-in hours for team members added pressure to the job in that team scores went down if log-in targets were not maintained. He said, "Most people coming in are not thinking of this as a career, only a stop gap. [However,] they end up going to another call center even if their plan is to get out of this." Seema, a twenty-five-year-old CSR, added that a good team was not easy to maintain. The call abandon rate for every 1,000 calls should not be more than 3 percent, a tall order if team members were not competent.

Anand, a twenty-seven-year-old married CSR, said he just could not get used to the changed rhythm. He listed several aspects of the job that caused him to seriously rethink this line of work:

> My shift is like night 9:00 p.m. to morning 10:00 a.m. By the time we come back it is afternoon or midday, so we can't even go to sleep ... Also we don't have our own PCs or cabins, you can log into any computer and its hard to not have your own space. The food is from the company; it is not like home food. See, if [the job has] a day timing, no... my family can cook and give it to me. The odd times at night—they can't get up at 11:00 p.m. or 12:00 p.m. in the night to cook for me and give it to me. And one more thing is the transportation. It [ITPL] is quite far off. This ITPL is out of town so there are a lot of problems—it (takes) me two hours to get to work and get back from work. I am losing sleep time. Totally four hours of transport apart from work.

Although various ring roads around the city provided easier navigation, many of these roads were either heavily congested with traffic or simply blocked because of further construction or deterioration. As gleaming call centers rise up on the outskirts of Bangalore, infrastructure in terms of roads and telecommunications struggles to keep up, and rural populations suddenly find themselves part of an expanding urban environment. Many farmers have sold their land to urban developers and bought tempos and Maruti vans to launch unlicensed transport services from the outlying rural areas to the city. The result is not just easier mobility for the labor population to and from the city but a significant increase in accidents and traffic jams along the outer ring roads.

Many respondents said they were deprived of a personal life and the job was maddeningly monotonous. The lack of physical contact with clients contributed to a sense of the surreal, where employees felt suspended between worlds, communicating only through voice and accent. The frequent changes in weekly work schedule made it difficult to plan personal events. Moreover, team managers such as Sneha, a twenty-six-year-old, had to stay on for up to "fifteen hours at a time if there is a shortage of staff, if there are unexpected calls to the assistant manager of operations, or if I have to fill for someone else."

Call center employees were also targets of racism. Theresa, a thirty-year-old CSR, said she was unprepared for the hateful comments some of her clients made to her. Shailaja, a twenty-five-year-old team manager, said that many of her CSRs reported that as soon as clients recognized the trace of an Indian accent, they would use the call to vent their frustrations at the outsourcing market. Clients in the United States who had lost their jobs to the industry would take it out on call center employees and berate them for being opportunistic and greedy. Recounting a personal experience, she said, "There are times when people say 'I want to speak to someone who speaks English,' or, 'I don't want to speak to an Indian.' I am seething with anger but I have to continue the call. There are some who are so nasty, especially some guys from the U.S." These caustic interactions were a great source of stress, as were other work-related ailments such as backaches, eyestrain and headaches, and indigestion (also see Singh and Pandey, 2005). Several team managers stated that stress in the workplace had resulted in more and more women smoking. Romantic and sexual relationships between team members were increasingly common. While male respondents discussed their jobs as transient, in that they would stay on until a better position came along, female respondents stated that they would work at the call center until they got married, or, if married, until their husband's income was stabilized.

From the discussion of employee experiences it is obvious that call center work is complex, involving shifting identities, economic benefits, and physical strain. Analysis of "the ground" of the industry also requires a discussion of its relationship to structure to arrive at an understanding of the long-term implications of call center transformations.

Why Can't We All Just Get Along?

In one breath at the end of his column on the potential of technology to expand the global economy, Friedman states (2006):

> If more countries can get just a few basic things right—enough telecom and bandwidth so their people can get connected; steadily improving education and decent, corruption-free economic governance; and the rule of law—and we can find more sources of clean energy, there is every reason for optimism that we could see even faster global growth in this century, with many more people lifted out of poverty. (p. 3)

If only. Colonial histories and entrenched cultural hierarchies in central and peripheral communities aside, technology will save the world. Certainly there is evidence for such optimism. The United States accounts for 59 percent of total global investment in the Indian ITES-BPO industry, targeting legal, logistics, and customer care segments. Europe is the second-largest market, at 22 percent, targeting HR, purchasing, finance, and accounting. Finally, the Asia/Pacific region follows at 15 percent, with the fastest-growing areas including HR, engineering, finance, accounting, and purchasing. The significant difference in costs—where overall operating, personnel, and property rental costs in India are just 20 percent, 14 percent, and 22 percent, respectively, those of the United States—makes estimates of annual growth rates as high as 200–300 percent not at all unlikely (NASSCOM, 2003). With such projections, many may argue that call centers provide an opportunity for rapid upward mobility and a quality of life for employees that would be impossible if they were to be employed by indigenous private and government institutions. Yet economic gains alone cannot tell the whole story; we have to consider the cultural implications of call center transformations for a postcolonial population.

We have to recognize, drawing from postcolonial theory, that "the world is an integrated ensemble of historical and regional processes, and that particular times and places can rarely be separated out from larger patterns if we are to make interpretations capable of producing change" (Schwarz, 2000, 5). Although Said's (1979) *Orientalism* has been critiqued extensively for its dichotomic analysis of the West and the East, it offers a good framework to understand some contemporary transnational processes. The consensual and flexible superiority of Orientalism, "which puts the Westerner in a whole series of possible relationships with the Orient without ever losing him the relative upperhand" (p. 25), carries a depressing relevance for the current context. Call centers provide ample evidence of the meticulous gentrification strategies by which natives are groomed to serve the MNCs and transnational bourgeoisie, many located in former imperial centers. Such grooming is facilitated by various "comprador intelligentsia" (Appiah, 1996, p. 62) who have been educated in English-language schools in mostly former British cantonment areas, perpetuating a structure of dominance and subordination set up centuries ago. Venn (2006) proposes *Occidentalism* as a process integrally intertwined with globalization. It includes the idea of modernity as a project to be realized, rational capitalism as a mode of production where subjects are transformed into commodities, and European colonialism as a mindset that places the "Other" as fundamentally inferior. The effects of Occidentalism are widespread and alive across the material world. Enduring structures of exploitation in the form of globalization and neocolonialism ensure that development or modernization, no matter how lucrative in the short term, continues to be uneven, as a colonial legacy, facilitating the continuation of cultural and media imperialism. Transactions continue according to a wide range of possibilities, without the elites in this structure ever losing the upper hand.

This study has provided a view into the experiences of call center employees who recognize the monetary benefits of their work, yet understand the toll it takes on their social and personal lives. Their statements on the social benefits of their work, where they learn to speak and communicate better in English, have to be critiqued as components of a larger ideological system that places high economic value on the language, accent, and behavior of the dominant class. For example, team leaders are keenly aware of the limits of their resistance because their very existence depends on complicity. As Butler (1997) points out, "how is survival to be maintained if the terms by which existence is guaranteed are precisely those that demand and institute subordination?" (p. 27). There is some room for change in that the health of the organization depends to a certain extent on employee well-being. Despite plentiful supply, recruiting, hiring, and training procedures are expensive and high turnover rates can spell doom for any organization in the long term. Recognizing these conditions, team leaders can, even though prevented from unionization, demand a certain standard in their work conditions. Yet they accept the limits of these demands and, in turn, impose the same on their teams, demanding long work hours, high call response and sales rates, and decreased customer complaints. It would be naïve, therefore, to assume that agency is always in opposition to power. Gramsci's notion of hegemony as a system of domination that induces consent in its subjects through ideology is particularly relevant here. Strinati (1995) explains:

> It can be argued that Gramsci's theory suggests that subordinated groups accept the ideas, values and leadership of the dominant group not because they are physically or mentally induced to do so, nor because they are ideologically indoctrinated, but because they have reason of their own. (p. 166)

Call center employees work within this ideology, accepting their cultural subordination as part of the job. As demonstrated in an earlier study of sweat shop laborers (McMillin, 2003), workers are aware of the drawbacks of the system but also recognize that equally lucrative alternatives are almost nonexistent. So it is not a reductionist matter of the material base defining culture, but of complex micro-negotiations within individual workers willing to transform what they consider their outside material sphere to sustain their inner spiritual sphere. The outsourcing of labor is integrally connected to the outsourcing of identity, the latter as much a commodity as the former in transnational flows of culture and capital, leading to dynamic and protean combinations of global and local, always in construction, never complete.

Notes

An earlier version of this paper, titled "Outsourcing Identities: Call Centers and Cultural Transformation in India," appeared in the Winter 2006 issue of *Economic and Political Weekly*, *41(3)*, 235–241.

1. The actual names of the call centers and the names of all respondents are confidential. Aliases are used where respondent names are provided in this chapter.
2. At the time of fieldwork, $1 was worth Rs. 45.

CONCLUSION

Moving On, Re-mixing It Up: Web 2.0, Offline/Online Intersections, Globalization through NGOs, Machinima, Mash-ups...

Radhika Gajjala

WHAT SORTS of convergences, conjunctures, and connections emerge in relation to globalization and migrant populations the world over as we move into what is being referred to as "Web 2.0"? Are "diaspora" and "globalization" still defined through particular lenses of privilege and lack, center and margin? With almost everyone on the Internet now claiming to be a marginalized voice being "empowered" to speak back to the center, where and how are we to locate the center? It becomes more important than ever, therefore, for Internet and cyberculture researchers to reexamine conceptual categories and frameworks such as "diaspora," "globalization," "new media," and even "empowerment," "multiculturalism," and "voice/voicelessness."

Researchers examining presence and absence or voice and voicelessness are mobilized to speak of identities emerging online, while binaries such as embodied/disembodied and global/local are deployed unproblematically in both utopian and dystopian viewpoints regarding the Internet. Performativity begins to shape exposure and privacy (see http://www.idtrail.org/content/section/11/95/ for examples of how these connections are being made). Thus, while claims are being made that the Internet is a "public sphere" in a Habermasian sense (Poster, 1995), corporate privation and surveillance come upon us in Internet-mediated environments and we learn to negotiate our speaking within interstices of presences and absences, cooperation and isolation, community engagement and individual consumerism. Simultaneously, hegemonic structures invested in particular ideologies of globalization and "free" markets learn to co-opt diverse identities and voices. Voice thus becomes a strategic construct in both cases.

In *Pedagogies of the Global*, the editor Arif Dirlik (2006) writes:

> Rather than erase difference by converting all to Euro/American norms of modernity, however, capitalist modernity, as it has gone global, has empowered societies once theoretically condemned to premodernity or tradition to make their own claims on modernity on the basis of those very traditions, as filtered through experiences of colonialism, neocolonialism, or simple marginalization by the forces of globalization. (p. 3)

Digital media play a significant role in aiding these connections and shaping these re-presentations. As an example, we can look at the new languages in which women's emancipation is articulated in global feminist communication/media networks.

On the basis of my research over the years examining Internet spaces in relation to offline community practice at the intersection of development and globalization, rural and urban, and how they are being impacted by multinational drives toward economic globalization, I want to suggest that these languages of women's emancipation in globalized media spaces are in fact recodings of familiar liberal feminist discourses interweaved with a capitalist, consumerist rhetoric of individual choice.

What I suggest is nothing new. Lexicons of women's empowerment and the "new" of media merge to perform a multicultural, "inclusive" global village while routing around communities whose image and praxis cannot be easily appropriated into the rhetoric of individual choice in consumption. The communities thus made invisible are often those that are part of hidden labor forces (migrant farm labor, sex workers, sweat shop labor, workers producing computer hardware, and so on), on whose social and economic immobility and lack of access to resources rest the mobility, individual-appearing choices, and freedom of other layers of the workforce who contribute to and benefit from multinational economic globalization.

This is being demonstrated in the way, for instance, NGOs are being increasingly deployed in a variety of ways to work on the intersections that traditional political and economic structures are unable (or unwilling) to engage in. NGOs appear in a broad continuum from complicity to resistance (or vice versa) in relation to what Dirlik (2006) refers to as capitalist modernity. Small-scale and large-scale funding agencies and plans have emerged (often as satellites to existing multinationals or governmental agencies) in the past fifteen to twenty years to provide support for nonprofits that will do the work that governments and corporations are unable to engage in for a variety of structural reasons.

Thus, these nonprofits begin to work in a model somewhat similar to BPOs and other small IT ventures funded through venture capitalists. The negatives of these sorts of movement include the erosion of benefits and government accountability (NGOs are put in the position of negotiating with policymakers as well as with the groups they seek to represent).

Global communities do not exist only in imaginary space or only online, without offline place-based linkages. These linkages are often made through economic activities that shape and are shaped by cultural practices and the workings of intergroup social power and politics in everyday life. Various local community histories and evolving contexts and the ways in which they negotiate (or renegotiate) connections and linkages around common interests, common profit, and common goals produce accountabilities and responsibilities that are central to global formations. What role do South Asian digital diasporas play in such formations? When and how are they both local and global, and in what ways does the dissemination across contexts influence and shift various consumer and producer cultures? Why does this matter? What is at stake? And is this at all "new"?

What is virtuality in today's context, then? I would argue that virtuality goes beyond digital mediation. The disconnect offered up by the simulacra of multiculturalism on the online stage is enacted through layered negations of strategic constructions of voice in the midst of "prodiences[1]" in Web 2.0 spaces where both the margin and center struggle to gain ground for representation, advertisement, and participation in globality. Thus we also need seriously to examine social networking systems (where the younger generations "hang out"), mash-ups, and machinima[2] from a first-world based celebration of networking and creativity, but also to understand how categories of race, caste, class, gender, and ethnicity are mobilized in a dialectical engagement—both complicit and resistant in varying degrees—with current trends in global capitalism, marketing, and consumerism.

"Old" and "new" mediated spaces do continue to impact online/offline networks. For, as Linda Leung argues in Chapter 1, the Internet should be viewed "as part of a wider media landscape, as a technology which borrows, hybridizes and diversifies from its predecessors." Thus practices and traditions from previous mediascapes do indeed shape our practices in relation to online multimediascapes. What is different, online, is how the interface, accessibility, and use of digital technologies such as the Internet reconfigure these ideologies and issues in their global/local reach.

This collection is a continuation of something begun in a special issue of *New Media and Society*; it started with trying to understand how South Asian diasporas are manifested online and at intersections of the global and the local. In thinking about putting together a collection of academic writings on South Asian digital diasporas, we found that economics and culture were inseparable, as traveling, migrating South Asians were incorporating online social networks into their everyday practices of living. Offline traditions such as "arranging" marriages, sharing of regional and vernacular music, remixing of music and film, and group religious information sharing were being replayed and shifted online as families and geographically connected groups of people became dispersed as a result of economically driven migration. In the case of diasporic South Asians, attempts at reconnecting and forming community by reconnecting dispersed family members and utilizing online technologies for cultural and religious practices shift the ways in which these practices shape their lives and society as a whole, impacting identity formations and politics offline.

Thus, as Rai (1995) and others have pointed out in past work in this area, online formations of nonresident Indians actually impacted events and politics in India in the early 1990s, when Hindu fundamentalist movements were on the rise. Offline cultural and religious practices for community support and discussion lead to material care being rebuilt in various settings. Furthermore, as several of the authors in this collection note, Internet technologies are enabling transnational labor flow of a different sort, allowing for skill sets to be developed differently and across time zones and cultures while at the same time privileging particular Western structures of economy, language, business practices, and sociocultural norms of communication and social interaction.

Thus there is an attempted "standardization" of bodies and business practices, and multicultural labor of various types is transmitted into multinational businesses whose organizational bodies are themselves spread across the globe geographically. Voices (as in call center) are literally trained into saleable accents; thus "voice" is clearly seen as a strategic construct for specific economic activities. Businesses are able to cut costs by paying less for the performed voices than they would have to for the "authentic" voices here in the United States—bypassing wage structures, etc., that would require the employer to invest more in hiring skilled labor in the U.S.

Much of this labor comes from regions in South Asia and China. As yet these voices are not "dis-"embodied, for the voicings come from bodies negotiating survival in space, economy, and culture—placing themselves at the (global) computer interface in relation to contexts offered up through multinational corporate structures. Therefore, their embodied travel is a bodily shifting that is no doubt significantly different from the embodied travel of immigration that physically relocates bodies across legal and geographical borders and territory. We had to rethink embodiment and movement across contexts. Thus we continued to ask:

> What kind of migratory subjects emerge in transnational spaces, at the intersection of the local and the global? What "regulatory fictions" and theoretical frames discipline (in a Foucauldian sense) manifestations of identity formations and communities online? Further, how do we perform our embodiment in cyberspace? What literacies are demanded in the performance of cyberbodies? What bodies are allowed embodiment through technologies? In turn, when viewed at the intersection of digital media and globalization, what kinds of bodies produce what kinds of technologies? What are the sociocultural transformations demanded in the name of "technological literacy" and underlying ideologies? (Gajjala, 2006, p. 179)

Thus, in preparing this collection, my coeditor and I began to see clearly how culture and economics are interwoven in the forming of digital diasporas. Digitally mediated technospaces are global. While globalization itself is nothing new—and the politics of mobilizing globalization and immigration as "new" must also be carefully examined—our task in this book was to examine how globalization in the digital age is happening and how South Asians continue to participate in this process at online/offline intersections.

We were able to see how process played out in various contexts. Therefore, while we see clear hierarchies that privilege particular literacies, we also see through the reading of the chapters in this book that while transnational capital both shapes and permits the existence of certain kinds of diversity in globalization, the multiplicities of identity formations in online settings suggest that globalization can never be completely subsumed by one particular dominant framework.

There is a need for more empirical work examining offline contexts through which the online is encountered and negotiated. Such work is being done in other contexts by researchers examining how teens in the United States encounter online environments (Scheidt, 2006) and how computer gamers engage virtual economies (Castranova, 2006; Dibbell, 2006; Taylor, 2006). The research sometimes starts online and then moves to investigating how the online becomes possible through offline socioeconomic practices. For instance, in the case of the work by Dibbell (2006), Taylor (2006), and Castranova (2005), each of them starts out with "in-world" ethnographies and analyses and moves to offline explorations. Other times the research begins offline with the examination of communities offline leading to examination of their online practices through cyberethnographic investigations and partnership action research projects and digital literacy projects (e.g., projects by organizations such as globalkids.org).

Such examinations will allow us to unpack how globalization is manifested across contexts. Future work in this area, therefore, needs to take seriously the study of offline/online intersections in diverse parts of the world. Several chapters in this book examine spaces of globalization, disembodiment, and identities. The next step would be to look at the *places* rooted in the physicality of urban and rural territories as various forms of economic and cultural embodied negotiations occur at and in relation to the computer interface (even when a computer is not physically present in that location, or when particular offline bodies have no direct presence in online space).

Critical ethnographic studies examining the relationship between online and offline contexts over a period of time are necessary if we are to provide an overall picture of how digital environments function in relation to processes of globalization. What we are also missing in this collection is an examination of vernacular diasporas, queer diasporas, and cross-generational diasporas. Such studies exist—either as works in progress or as very recent publications. For instance, in *Mobile Cultures: New Media in Queer Asia*, Berry, Martin, and Yue (2003) point to some of the queer possibilities inherent in new discourses and practices of transnational Asian and Internet regionalism, which are themselves responses to globalization.

Such examinations of the interplay between globalization and practices mediated by transnational travel, work, and play, make it possible to observe how globalization processes do not merely reinstate older hierarchies of gender, sexuality, race, and class. While transnational capital both shapes and permits the existence of certain kinds of diversity in globalization, these same multiplicities of identity formations that allow us to be performers on the global stage in particular contained ways also sometimes suggest that globalization can never be completely subsumed by one particular dominant framework. In the emerging connections between queer identities and online work and play, transgressions are visible when we examine how specific intersections are inhabited by generations more familiar with and comfortable in such environments (i.e., those who have grown up taking the Internet and the World Wide Web and computer games as a fact of daily life).

The chapters in this book provide examples of how South Asian digital diasporas have been studied at the intersection of culture and economics. By no means is it our claim that the collection is comprehensive. In fact, it would be impossible for any collection on this topic to make such a claim—even as this book goes into production, new forms of online/offline engagement are emerging as the younger generation compels us to use Web 2.0 technologies.

Businesses, educators, and social and political organizations are migrating into these formats. Social network systems such as Facebook, MySpace, Orkut, and Hi5, as well as three-dimensional social environments such as *Second Life*, are inhabited by large numbers of younger generation users ("digital natives"). Digital divides are taking on nuances not limited to issues of access to hardware and software. Diasporic South Asians and their Bollywood fan networks are busy setting up networks that intersect with film, television, music, and software programming.

At the same time as Globalization through model diasporas is occurring, NGOs, as well as historically marginalized groups around the world, are asserting their voices in relation to the global economy. Antiglobalization activists also occupy Internet space. How are all these received, and what techniques and strategies are employed in production to target and lure audiences? Advertising and usability research tends not to address the politics of marketing in these Internet-mediated environments while naturalizing particular kinds of consumption. Similarly, critical theorists rarely enter into explorations of the practices of marketing in these spaces, even as the consumption and consumer cultures are studied all the time. Online, in the case of "prodiences," there are opportunities to engage these intersections in immersive environments cyberethnographically as corporations and nonprofits all rush to set up shop at every new interface that emerges.

Notes

1. This was one of the "four Ps" that Toby Miller talked about at a panel on global media studies at the International Communication Association conference held in San Francisco in May 2007. The four Ps he mentioned are Pirates, PredItors (simultaneously producers and editors), Prodiences (audiences who are also producers), and Polluters.
2. According to the wikipedia (see http://en.wikipedia.org/wiki/Machinima), "Machinima is an example of emergent gameplay, a process of putting game tools to unexpected ends, and of artistic computer game modification. The real-time nature of machinima means that established techniques from traditional film-making can be reapplied in a virtual environment."

REFERENCES

Introduction

Enteen, J. (2006). Spatial conceptions of URLs: Tamil Eelam networks on the world wide web. *New Media & Society, 8,* 229–249.

Jeganathan, P. (1998). Eelam.com: Place, nation and imagi-nation in cyberspace. *Public Culture, 10(3),* 515–528.

Chapter One

Apple, M. (1992). Is the new technology part of the solution or part of the problem in education. In J. Beynon & H. Mackay (Eds.), *Technological literacy and the curriculum* (pp. 105–124). London: Falmer.

Brah, A. (1992). Differences, diversity and differentiation. In J. Donald & A. Rattansi (Eds.), *"Race," culture and difference* (pp. 126–147). London: Sage.

ClickZ Stats. (2001). Asians among most wired Americans. Retrieved March 22, 2004, from http://www.clickz.com/stats/big_picture/demographics/article.php/5901_942621.

ClickZ Stats. (2006a). Top 15 online populations and Web properties worldwide, June 2006. Retrieved September 11, 2006 from http://www.clickz.com/showPage.html?page=3623166.

ClickZ Stats. (2006b). Population explosion! Retrieved September 11, 2006, from http://www.clickz.com/showPage.html?page=151151.

Fernback, J. (1999). There is a there there: Notes towards a definition of cybercommunity. In S. Jones (Ed.), *Doing Internet research* (pp. 203–220). London: Sage.

Gajjala, R. (1999). Cyborg diaspora and virtual imagined community: Studying SAWNET. *Cybersociology Research Methodology, 6.* Retrieved June 4, 2004, from http://www.cybersociology.com.

Giddens, A. (1991). *Modernity and self-identity: Self and society in the late modern age.* Cambridge: Polity.

Gilroy, P. (1990/1991). It ain't where you're from, its where you're at...the dialectics of diasporic identification. *Third Text, 13,* 3–16.

Greenspan, R. (2002). China pulls ahead of Japan. Retrieved April 15, 2004, from http://www.clickz.com/stats/sectors/geographics/article.php/1013841.

Greenspan, R. (2003a). African-Americans create online identity. Retrieved March 22, 2004, from http://www.clickz.com/stats/big_picture/demographics/article.php/3084241.

Greenspan, R. (2004). Creating, contributing content catching on. Retrieved March 22, 2004, from http://www.clickz.com/stats/big_picture/applications/article.php/3319651.

HMSO. (1991). *The 1991 census on CD-ROM.* Cambridge: Chadwyck-Healy.

Hoffman, D. & Novak, T. (1998). Bridging the digital divide: The impact of race on computer access and Internet use. Retrieved June 4, 2004, from http://elab.vanderbilt.edu/research/papers/html/manuscripts/race/science.html.

hooks, b. (1992). *Black looks*. London: Turnaround.

Interrogate the Internet. (1996). Contradictions in cyberspace: Collective response. In R. Shields (Ed.), *Cultures of Internet* (pp. 125–132). London: Sage.

Johnson, F. (1996). Cyberpunks in the White House. In J. Dovey (Ed.), *Fractal dreams: New media in social context* (pp. 78–108). London: Lawrence & Wishart.

Jones, J. (1993). The construction of black sexuality: Towards normalizing the black cinematic experience. In M. Diawara (Ed.), *Black American cinema* (pp. 247–256). New York: Routledge.

Kendall, L. (1999). Recontextualising "cyberspace": Methodological considerations for online research. In S. Jones (Ed.), *Doing Internet research* (pp. 57–74). London: Sage.

Khan, F. (1988). The inner city context and the generation of curricula in adult education. In *SCUTREA 25th Anniversary CD-ROM Conference Proceedings (1970–1997).* London: Standing Conference on University Teaching and Research in the Education of Adults.

Lippard, L. (1992). Mapping. In F. Frascina & J. Harris (Eds.), *Art in modern culture* (pp. 160–169). London: Phaidon.

Mallapragada, M. (2000). The Indian diaspora in the USA and around the Web. In D. Gauntlett (Ed.), *Web studies* (pp. 179–185). London: Arnold.

Melkote, S., & Liu, D. (2000). The role of the Internet in forging a pluralistic integration: A study of Chinese intellectuals in the United States. *International Communication Gazette, 62(6),* 495–504.

Miniwatts Marketing Group. (2006). Internet usage in Asia. Retrieved September 11, 2006, from http://www.internetworldstats.com/stats3.htm.

Mitra, A. (1997). Virtual commonality: Looking for India on the Internet. In S. Jones (Ed.), *Virtual culture: Identity and communication in cybersociety* (pp. 55–79). London: Sage.

Mitra, A. & Cohen, E. (1999). Analyzing the Web: Directions and challenges. In S. Jones (Ed.), *Doing Internet research* (pp. 179–202). London: Sage.

Modood, T. (1997). "Difference," cultural racism and anti-racism. In P. Werbner, & T. Modood (Eds.), *Debating cultural hybridity: Multicultural identities and the politics of anti-racism* (pp. 154–172). London: Zed.

Morrissey, B. (2003). Black online population narrows adoption gap. Retrieved March 22, 2004, from http://www.clickz.com/stats/big_picture/demographics/article.php/1855931.

Nakamura, L. (2002). *Cybertypes: Race, ethnicity and identity on the Internet.* New York: Routledge.

Nelson, A. & Tu, T. (2001). *Technicolor: Race, technology and everyday life.* New York: New York University.

Panos. (1998). The Internet and poverty: Real help or real hype? Retrieved June 4, 2004, from http://www.panos.org.uk/resources/reportdetails.asp?id=1049.

Pastore, M. (2000). European Internet use still behind the US. Retrieved March 22, 2004, from http://www.clickz.com/stats/big_picture/geographics/article.php/5911_351591.

Pastore, M. (2001a). Internet remains a man's domain. Retrieved May 5, 2005, from http://www.clickz.com/stats/sectors/demographics/article.php/5901_809341.

Pastore, M. (2001b). British Internet market continues to grow, mature. Retrieved March 22, 2004, from http://www.clickz.com/stats/big_picture/geographics/article.php/5911_718651.

Rix, V. (1997). Industrial decline, economic restructuring and social exclusion in London East throughout the 1980s and 1990s. *Rising East: The Journal of East London Studies, 1(1)*, 118–141

Spender, D. (1996). *Nattering on the Net: Women, power and cyberspace.* North Melbourne: Spinifex.

Terranova, T. (1996). Digital Darwin: Nature, evolution and control in the rhetoric of electronic communication. *New Formations: TechnoScience, 29*, 69–83.

Wakeford, N. (2000). New media, new methodologies: Studying the Web. In D. Gauntlett (Ed.), *Web studies* (p. 31–42). London: Arnold.

Zurawski, N. (1996). Ethnicity and the Internet in a global society. Retrieved September 13, 1996, from http://www.uni-muenster.de/PeaCon/zurawski/inet96.html.

Chapter Two

Hobday, M. (1995). East Asian latecomer firms: Learning the technology of electronics. *World Development, 23(7)*, 1171–1193.

Jayaram, N. (1998). The study of Indian diaspora: A multidisciplinary agenda. Occasional papers, Center for the Study of Indian Diaspora, University of Hyderabad, India.

Kapur, D., & Ramamurti, R. (2001). India's emerging competitive advantage in services. *Academy of Management Executive, 15(2)*, 20–32.

Laxminarayan, K. (1998). Indian diaspora: A demographic perspective. Occasional papers, Center for the Study of Indian Diaspora, University of Hyderabad, India.

NASSCOM Strategic Review (Executive Summary) (2007). Retrieved January 31, 2007, from http://www.nasscom.in.

Portes, A. (1996). Global villagers: The rise of transnational communities. *American Prospect*, 7(25), 74–77.

Saxenian, A. (2002). *Local and global networks of immigrant professionals in Silicon Valley.* San Francisco: Public Policy Institute of California.

Saxenian, A., Bresnahan T., & Gambardella, A. (2001). "Old Economy" inputs for "New Economy" outcomes: Cluster formation in the new Silicon Valleys. *Industrial and Corporate Change, 10(4)*, 835–860.

Singhal, A. & Rogers, E. M. (2001). *India's communication revolution from bullock carts to cyber marts.* New Delhi: Sage.

Tinker, H. (1993). *A new system of slavery: The export of Indian labor overseas, 1830–1920.* 2nd ed. London: Hansib.

UNDP. (2001). *Human Development Report, 2001.* New York: United Nations Development Program.

UNDP. (2005). *Human Development Report, 2005.* New York: United Nations Development Program.

Chapter Three

Anderson, B. (1991). *Imagined communities.* London: Verso.
Bourdieu, P. (1984). *Distinction: A social critique of the judgment of taste.* London: Routledge.
Duncan, J., & Gregory, D. (1999). *Writes of passage: Reading Travel Writing.* Routledge: London and New York
Gajjala, R. (1998).The SAWnet refusal: An interrupted cyberethnography. Doctoral dissertation, University of Pittsburgh. Dissertation Abstracts International (UMI No. 9900131).
Gajjala, R. (2004). *Cyberselves: Feminist ethnographies of South Asian women.* Lanham, MD: Altamira Press.
Jones, S. G. (Ed.). (1995). *CyberSociety: Computer-mediated communication and community.* Thousand Oaks, CA: Sage.
Karamcheti, I. (1992). The shrinking Himalayas. *Diaspora, 2(2),* 261–276
Lal, V. (1999). The politics of history on the Internet: Cyber-diasporic Hinduism and the North American Hindu diaspora. *Diaspora, 8(2),* 137–172.
Mitra, A. (1997).Virtual commonality: Looking for India on the Internet. In S. Jones (Ed.), *Virtual culture: Identity and communication in cybersociety* (pp. 55–79). London: Sage.
Nelson, E. (Ed.). (1992). *Reworlding: The literature of the Indian diaspora.* New York: Greenwood.
Pratt, M. L. (1992). *Imperial eyes: Travel writing and transculturation.* New York: Routledge.
Rai, A. S. (1995). India on-line: Electronic bulletin boards and the construction of a diasporic Hindu identity. *Diaspora, 4(1),* 31–57.
Sadowski-Smith, C. U. S. (1999). Border theory, globalization, and ethnonationalisms in post-Wall Eastern Europe. *Diaspora, 8(1),* 3–22.
Slater, D., & Miller (2000). *The Internet: An ethnographic approach.* Oxford: Berg.
Spivak, G. C. (1999). *A critique of postcolonial reason: Toward a history of the vanishing present.* Cambridge, MA: Harvard University Press.
Stone, R. (1992). Virtual systems. In J. Crary & S. Kwinter (Eds.), *Incorporations* (pp. 609–621). New York, NY: Zone.
Watson, N. (1997). Why we argue about virtual community: A case study of the Phish. Net fan community. In S. Jones (Ed.), *Virtual culture: Identity and communication in cybersociety* (pp. 102–132). London: Sage.
Winner, L. (1997). Cyberlibertarian myths and the prospects for community. Retrieved February 18, 2005, from http://www.langdonwinner.org/index.html.

Chapter Four

Berry, E. M. (2003, November 26). The last hustle. *The Village Voice*. Retrieved April 13, 2006, from http://www.villagevoice.com/news/0348,berry,48963,1.html.

BMI. (2006). BMI forecasts U.S. ringtones sales to hit $600 million in 2006. Retrieved April 13, 2006, from http://bmi.com/news/200604/20060403a.asp.

Chadha, T. (2003, July 2–8). Mix this. *The Village Voice*. Retrieved April 13, 2006, from http://www.villagevoice.com/news/0327,chadha,45230,1.html.

Cooper, C. (2004). *Sound clash: Jamaican dancehall culture at large*. New York: Palgrave Macmillan.

Deleuze, G., & Guattari, F. (1987). *A thousand plateaus: Capitalism and schizophrenia* (B. Massumi, Trans.). Minneapolis: University of Minnesota Press.

Gilroy, P. (1993). *The black Atlantic: Modernity and double consciousness*. Cambridge: Harvard University Press.

Hemment, D. (2004). Affect and individuation in popular electronic music. In I. Buchanan & M. Swiboda (Eds.), *Deleuze and music* (pp. 77–94). Edinburgh: Edinburgh University Press.

Howard-Spink, S. (2005). Grey Tuesday, online cultural activism and the mash-up of music and politics. *First Monday, 9(10)*. Retrieved April 13, 2006, from http://www.firstmonday.org/issues/issue9_10/howard/.

Ilx. (2002). June 5, 2002. Retrieved April 13, 2006, from http://ilx.wh3rd.net/thread.php?msgidÿ2195552.

Jones, L. (1967). *Black music*. New York: Morrow Paperback Editions.

King, J. (2001). Rap and Feng Shui: On ass politics, cultural studies, and the Timbaland sound. In T. Miller (Ed.), *Companion to cultural studies* (pp. 430–453). Oxford: Blackwell Publishers.

Kraidy, M. (2005). *Hybridity, or the cultural logic of globalization*. Philadelphia: Temple University Press.

Latour, B. (1999). On recalling ANT. In J. Law & J. Hansard (Eds.), *Actor network theory and after* (pp. 15–25). Oxford: Blackwell Publishers.

Maira, S. (2002). *Desis in the house: Indian American youth culture in New York City*. Philadelphia: Temple University Press.

Manuel, P. (1993). *Cassette culture: Popular music and technology in north India*. Chicago: University of Chicago Press.

Melwani, L. (2002). Desi Flava. Little India. Retrieved April 13, 2006, from http://www.littleindia.com/june2002/Desi%20Flava.htm.

Miller, K. (2004). Bolly-hood remix. *Newsletter, 33(2)*. Retrieved April 13, 2006, from the Institute for Studies in American Music, Conservatory of Music, Brooklyn College of the City University of New York Web site: http://depthome.brooklyn.cuny.edu/isam/S04Newshtml/Bollyhood/Bollyhood.htm.

Mitchell, T. (2001). *Global noise: Rap and hip-hop outside the USA*. Middletown, CT: Wesleyan University Press.

Oswald, J. (2004). Bettered by the borrower: The ethics of musical debt. In C. Cox & D. Warner (Eds.), *Audio culture: Readings in modern music* (pp. 131–137). New York & London: Continuum Books.

Prashad, V. (2000). *The karma of brown folk.* Minneapolis: University of Minnesota Press.

Radano, R. (2003). *Lying up a nation: Race and black music.* Chicago: University of Chicago Press.

Reynolds, S. (2004). Lost in music: Obsessive record collecting. In E. Weisbard (Ed.), *This is pop: In search of the elusive at Experience Music Project* (pp 289–307). Cambridge, MA: Harvard University Press.

Schloss, J. (2004). *Making beats: The art of sample-based hip-hop.* Middletown, CT: Wesleyan University Press.

Sharma, N. (2005). Musical crossings: Identity formations of second-generation South Asian American hip hop artists. Institute for the study of social change fellows working papers. Retrieved April 13, 2006, from University of California, Berkeley Web site http://repositories.cdlib.org/issc/fwp/ISSC_WP_02/.

Sharma, S. (2003). The sounds of alterity. In M. Bull & L. Back (Eds.), *The auditory culture reader* (pp. 409–418). Oxford: Berg.

Smith, C. H. (2003). "I don't like to dream about getting paid": Representations of social mobility and the emergence of the hip-hop mogul. *Social Text, 21(4)*, 69–97.

Taylor, T. (2004). Some versions of difference: Discourses of hybridity in transnational musics. In T. G. Oren & P. Petro (Eds.), *Global currents: Media and technology now* (pp. 219–244). Piscataway, NJ: Rutgers University Press.

Weheliye, A. G. (2002). "Feenin": Posthuman voices in contemporary black popular music. *Social Text, 20(2)*, 21–47.

Wolk, D. (2004). Compressing pop: How your favorite song got squished. In E. Weisbard (Ed.), *This is pop: In search of the elusive at Experience Music Project* (pp. 212–222). Cambridge, MA & London: Harvard University Press.

Zuberi, N. (2002). India song: Popular music genres since economic liberalization. In D. Hesmondhalgh & K. Negus (Eds.), *Popular music studies* (pp. 238–250). London: Arnold/New York: Oxford University Press.

Chapter Five

Arnold A. (1993). Aspects of production and consumption in the popular Hindi film song industry. *Asian Music, 24(1)*, 122–136.

Bakshi, R. (1998). Raj Kapoor. In A. Nandy (Ed.), *The secret politics of our desires: Innocence, culpability and Indian popular cinema* (pp. 92–133). New Delhi: Zed Books.

Benjamin, W. (1999). *The Arcades project* (H. Eiland & K. McLaughlin, Trans.). Cambridge, MA: Belknap Press.

Benjamin, W. (2003). Paris, the capital of the nineteenth century. In *Selected Writings* (Vol. 3) (H. Eiland, Trans.) (pp. 40–41). Cambridge, MA: Belknap Press.

Black. (2005, June 19). Something "special" in Bollywood. *Sunday Herald Entertainment* Retrieved October 1, 2005, from http://www.deccanherald.com/deccanherald/jun192005/enter131232005617.asp.

Bordwell, D. (2002). Intensified continuity: Visual style in contemporary American film. *Film Quarterly, 55(3)*, 16–28.

Chakravarty, S. S. (1993). *National identity in Indian popular cinema 1947–1987.* Austin: University of Texas Press.

Chute, D. (2002). Bollywood rising: A beginner's guide to Hindi cinema. *Film Comment, 38(3)*, 35–40.

Deleuze, G. (2005). *Cinema 2: The time image and Francis Bacon: The logic of sensation* (D. Smith, Trans.). Minneapolis, MN: University of Minnesota Press.

Deleuze, J., & Parnet, C. (1987). *Dialogues.* London: Athlone.

Derrida, J. (1982). *Positions* (A. Bass, Trans.). Chicago: University of Chicago Press.

Doctorow, C. (2005). 0wnz0red. In J. Dann (Ed.), *Nebula awards showcase 2005* (pp. 106–130). New York: Penguin.

Ford, J. (2002). The empire of ice cream. In J. Dann (Ed.), *Nebula Awards Showcase 2005* (pp. 254 –270). New York: Penguin.

Ghosh, L. B. (2005, September 6). Bollywood goes all out for special effects. *The Hindu.* Retrieved October 1, 2005, from http://www.hindu.com/2005/09/06/stories/2005090607090200.htm.

Gopalan, L. (2002). *Cinema of interruptions.* London: British Film Institute.

Guzzetti, A. (1975). Narrative and the film image. *New Literary History, 6(2)*, 379–392.

Jaan Production. (2000). Dilip Kumar: Bollywood's allstar. Retrieved October 1, 2005, from http://www.pyara.com/dilip.htm.

Jalal, A. (2002, November 2–9). A Letter to India: In Manṭo's spirit. *EPW.* Retrieved December 1, 2005, from http://www.epw.org.in/showArticles.php?root=2002&leaf=11&filename=5119&filetype=html.

Jayaprakash, Y. T. (2000). Remote audiences beyond 2000: Radio, everyday life and development in South India. *International Journal of Cultural Studies, 3(2)*, 227–239.

Jha, P. (2003). Lyrical nationalism: Gender, friendship, and excess in 1970s Hindi cinema. *Velvet Light Trap, 51*, 43–53.

Kehr, D. (2002, December 20). Shot in Los Angeles, but Bombay all the way. *New York Times*, p. E32.

Marks, L. (1999). *The skin of the film: Intercultural cinema, embodiment, and the senses.* Durham, NC: Duke University Press.

Marks, L. (2002). *Touch: Sensuous theory and multisensory media.* Minneapolis, MN: University of Minnesota Press.

Marx, K. (1867). *Capital* (Vol. 1, Pt. IV, chap. 14). Retrieved August 1, 2006, from http://www.marxists.org/archive/marx/works/cw/volume35/index.htm.

Massumi, B. (2001). *Parables for the virtual.* Durham, NC: Duke University Press.

Mishra, V. (2001). *Bollywood cinema: Temples of desire.* New York: Routledge.

Morson, G. S. (2003). Narrativeness. *New Literary History, 34,* 59–73.

Niyogi De, E. (2002). Modern Shakespeares in popular Bombay cinema. *Screen, 43(1),* 19–40.

Noronha, F. (2004, December 4.). Censors continue to stifle Indian filmmaking. *Indo-Asian News Service.* Retrieved November 1, 2005, from http://in.movies.yahoo.com/041204/43/2i9o8.html.

Parisi, L. (2004). Information trading and symbiotic micropolitics. *Social Text, 22(3),* 25 –49.

Pashupati, K., Sun, H. L., & McDowell, S. D. (2003). Guardians of culture, development communicators, or state capitalists? A comparative analysis of Indian and Chinese policy responses to broadcast, cable and satellite television. *Gazette, 65(3),* 251–271.

Pillai, S. (2003, June 19.). Hollywood meets Bollywood. *The Hindu.* Retrieved October 1, 2006, from http://www.hindu.com/thehindu/mp/2003/06/19/stories/2003061900110203.htm.

Prasad, M. M. (2000). *Ideology of the Hindi film: A historical construction.* New York: Oxford University Press.

Rajagopal, A. (1993). The rise of national programming: The case of Indian television. *Media, Culture and Society, 15(1),* 91–111.

Rajagopal, A. (2001). *Politics after television: Hindu nationalism and the reshaping of the public in India.* Cambridge: Cambridge University Press.

Ramakrishna, B. (2003). Smarter than the smart: Cinema marketing. *Theatre World, 59,* 19–22.

Rushton, R. (2002). What can a face do? On Deleuze and faces. *Cultural Critique, 51,* 219–237.

Smith, D. (2005). Deleuze on Bacon: Three conceptual trajectories in *The Logic of Sensation.* In G. Deleuze, *Francis Bacon: The logic of sensation* (pp. vii–xxvii) (D. Smith, Trans.). Minneapolis, MN: University of Minnesota Press.

Sonwalkar, P. (2001). India: Makings of little cultural/media imperialism? *Gazette, 63(6),* 505–519.

Star Profile: Suchitra Sen. (1955, August 5). *Filmfare,* pp. 12–13.

Thomas, P. N. (1999). Trading the nation. *International Communication Gazette, 61(3–4),* 275–292.

Thompson, K. (1986). The concept of cinematic excess. In P. Rosen (Ed.), *Narrative, apparatus, ideology* (pp. 130–142). New York: Columbia University Press.

Vasudevan, R. S. (1995). Addressing the spectator of the "third world" national cinema: The Bombay "social" film of the 1940s and 1950s. *Screen, 36(4),* 305–324.

Webopedia. (2004). MPEG. Retrieved February 21, 2006, from http://www.webopedia.com/TERM/M/MPEG.html.

Chapter Six

Aloysius, G. (2003). Poverty and development in India: A sociological consideration. Unpublished manuscript, Jamia Milia University, Delhi.

Anand, S. (2005). Covering caste: Visible Dalit, invisible Brahman. In R. Nalini (Ed.), *Practising journalism: Values, constraints, implications* (pp. 172–198) Thousand Oaks, CA: Sage.

Anderson, B. (1983). *Imagined communities: Reflections on the origin and spread of nationalism*. New York: Verso.

Basu, D. (1987). *Introduction to constitution of India*. New Delhi: Prentice Hall.

Chatterjee, P. (1992). *Nationalist thought and the colonial world: A derivative discourse*. Delhi: Oxford University Press.

Curran, J. (1991). Rethinking the media as the public sphere. In P. Dahlgren & C. Sparks (Eds.), *Communication and Citizenship: Journalism and the Public Sphere* (pp. 27–57). London: Routledge.

Derrida, J. (1976). *Of grammatology*. (G. C. Spivak, Trans.). London: Johns Hopkins University Press.

Derrida, J., & Stiegler, B. (2002). *Echographies of television*. London: Polity.

Gandhi, M. K. (1938). *Hind Swaraj*. Ahmedabad: Navjeevan.

Guru, G. (2001). Politics of representation. *Seminar*, *38* [Electronic version].

Guru, G. (2004). The Language of Dalitbahujan political discourse. In M. Mohanty (Ed.), *Class, caste, gender* (pp. 256–269). Delhi: Sage.

Jeffrey, R. (2003). *India's newspaper revolution: Capitalism, politics and the Indian-language press*. New Delhi: Oxford University Press.

Jeffrey, R. (2000). *The newspaper revolution*. Delhi: Sage.

Kaviraj, S. (1993). The imaginary institution of India. In P. Chatterjee & G. Pandey (Eds.), *Subaltern studies* (Vol. VII) (pp. 1–39). Delhi: Oxford University Press.

Kothari, R. (1993). *Poverty: Human consciousness and the amnesia of development*. London: Zed Books.

McLuhan, M. (1964). *Understanding media*. New York: McGraw Hill Book Company.

Nandy, A. (1983). *The intimate enemy: Loss and recovery of self under colonialism*. New Delhi: Oxford University Press.

Parekh, B. (1999). *Colonialism, tradition and reform: An analysis of Gandhi's political discourse*. New Delhi: Sage.

Parel, A. (1997). *Hind Swaraj and other writings*. New Delhi: Foundation Books.

Prasad, C. (2004). *The Dalit diary*. Chennai: Navyana.

Rege, S. (2000, February 5–11). "Real feminism" and Dalit women. *Economic and Political Weekly*. Retrieved from http://www.epw.org.in.

Watts, M. (2004). Antinomies of community: Some thoughts on geography, resources and empire [Electronic Version]. *Transactions of the Institute of British Geographers, 29(2)*, 195–216.

Chapter Seven

Appadurai, A. (1996). *Modernity at large: Cultural dimensions of globalization.* Minneapolis: University of Minnesota Press.

Azra. (2002). Diary entry. In R. Vasudevan, R. Sundaram, J. Bagchi, M. Narula, G. Lovink, & S. Sengupta (Eds.), *Sarai reader 02: The cities of everyday life.* New Delhi: Sarai, CSDS+ The Society for Old and New Media. Retrieved November 10, 2005, from http://www.sarai.net/publications/readers/02-the-cities-of-everyday-life.

Bagchi, J. (Ed.). (2002). Cybermohalla diaries. In R. Vasudevan, R. Sundaram, J. Bagchi, M. Narula, G. Lovink, & S. Sengupta (Eds.), *Sarai reader 02: The cities of everyday life* (pp. 176–191). New Delhi: Sarai, CSDS+ The Society for Old and New Media. Retrieved November 10, 2005, from http://www.sarai.net/publications/readers/02-the-cities-of-everyday-life

Blunt, A. (1994). Mapping authorship and authority: Reading Mary Kingsley's landscape descriptions. In A. Blunt & G. Rose (Eds.), *Writing women and space: Colonial and postcolonial geographies* (pp. 51–72). New York: Guilford.

Coleman, D., Danze, E., & Henderson, C. (1996). *Architecture and feminism.* New York: Princeton Architectural Press.

Galiyon se / By lanes. (2002). Sarai: The New Media Initiative A Programme of the Centre for the Study of Developing Societies. Ankur: Society for Alternatives in Education. Retrieved November 10, 2005, from http://www.sarai.net/practices/cybermohalla/public-dialogue/books/galiyon-se

Harvey, D. (1990). Between space and time: Reflections on the geographical imagination. *Annals of the Association of American Geographers, 80*, 418–434.

Harvey, D. (1996). *Justice, nature, and the geography of difference.* Oxford: Blackwell.

Hayden, D. (1997). *The grand domestic revolution: A history of feminist designs for American homes, neighborhoods, and cities.* Cambridge, MA: MIT Press.

hooks, b. (1984). *Feminist theory: From margin to center.* Boston: South End Press.

Kaup, M. (1997). The architecture of ethnicity in Chicano literature. *American Literature, 69*, 361–397.

Markusen, A. R. (1980). City spatial structure, women's household work, and national urban policy. *Signs, 5(3)*, S22–S44.

Rose, G. (1993). *Feminism and geography: The limits of geographical knowledge.* Minneapolis: University of Minnesota Press

Sassen, S. (1998). *Globalization and its discontents: Essays on the new mobility of people and money.* New York: New Press.

Smith, M. P. (1992). Postmodernism, urban ethnography, and the new social space of ethnic identity. *Theory and Society, 21*, 493–531.

Soja, E. (1996a). Afterword. *Stanford Law Review, 48,* 1421–1429.

Soja, E. (1996b). *Thirdspace: Journeys to Los Angeles and other real-and-imagined places.* Cambridge, MA: Blackwell.

Spain, D. (1993). Gendered spaces and women's status. *Sociological Theory, 11,* 137–151.

Tabassum, A., Ali, S., Rai, S., Neelofer, Ayesha, Shahjehan, Bobby, et al. (2002). Cybermohalla diaries. In R. Vasudevan, R. Sundaram, J. Bagchi, M. Narula, G. Lovink, & S. Sengupta (Eds.), *Sarai reader 02: The cities of everyday life* (pp. 176–190). New Delhi: Sarai, CSDS+ The Society for Old and New Media. Retrieved November 10, 2005, from http://www.sarai.net/publications/readers/02-the-cities-of-everyday-life.

Weedon, C. (1997). *Feminist practice and poststructuralist theory* (2nd ed). Oxford: Basil Blackwell.

Weedon, C. (1999). *Feminism, theory and the politics of difference.* Malden, MA: Blackwell.

Wilson, E. (1991). *The sphinx in the city: Urban life, control of disorder, and women.* Los Angeles: University of California Press.

Wyly, E. K. (1999). Continuity and change in the restless urban landscape. *Economic Geography, 75,* 309–338.

Chapter Eight

Brah, A. (1996). *Cartographies of diaspora: Contesting identities.* London: Routledge.

Chowdhury, P. (1996). Enforcing cultural codes: Gender and violence in northern India. In K. Jayawardena & M. de Alwis (Eds.), *Embodied violence: Communalising women's sexuality in South Asia* (pp. 332–367). New Delhi and London: Kali for Women.

Dirks, N. B. (2001). *Castes of mind: Colonialism and the making of modern India.* Princeton, NJ, and Oxford: Princeton University Press.

Habermas, J. (1993/1962). *The structural transformation of the public sphere: An inquiry into a category of bourgeois society* (T. Burger, Trans.). Cambridge, MA: MIT Press.

Hall, S. (1990). Cultural identity and diaspora. In J. Rutherford (Ed.), *Identity: Community, culture, difference* (pp. 222–237). London: Lawrence & Wishart.

Human Rights Watch. (1999). Broken people: Caste violence against India's untouchables. Retrieved March 31, 2006, from http://www.hrw.org/reports/1999/india/India994.htm.

Kolko, B. (2003). Introduction: The reality of virtuality. In B. Kolko (Ed.), *Virtual publics: Policy and community in an electronic age* (pp. 1–8). New York: Columbia University Press.

Lal, V. (1999). The politics of history on the Internet: Cyber-diasporic Hinduism and the North American Hindu diaspora. *Diaspora, 8(2)*, 22–23.

Lockard, J. (2000). Babel machines and electronic universalism. In B. E. Kolko, L. Nakamura & G.B. Rodman (Eds.), *Race in cyberspace* (pp.171–190). London: Routledge.

Mitra, A. (1997). Virtual commonality: Looking for India on the Internet. In S. Jones (Ed.), *Virtual culture: Identity and communication in cybersociety* (pp. 55–79). Thousand Oaks, CA: Sage.

Morley, D. (2000). *Home territories: Media, mobility and identity.* London: Routledge.

Ong, A. (2003). Cyberpublics and diaspora politics among transnational Chinese. *Interventions, 5(1)*, 82–100.

Paz, O. (1995). *In light of India.* San Diego, CA: Harcourt Brace.

Poster, M. (2001). Cyberdemocracy: The Internet and the public sphere. In D. Trend (Ed.), *Reading digital culture* (pp. 212–229). Oxford: Blackwell Publishers.

Rai, A. (1995). India on-line: Electronic bulletin boards and the construction of a diasporic Hindu identity. *Diaspora, 4(1)*, 31–57.

Raj, D. S. (2003). *Where are you from? Middle-class migrants in the modern world.* London: University of California Press.

Reeves, T., & Bennet, C. (2004). We the people: Asians in the United States. Census 2000 Special Reports. Retrieved December 17, 2005 from http://www.census.gov/prod/2004pubs/censr-17.pdf.

Saco, D. (2002). *Cybering democracy: Public space and the Internet.* Minneapolis: University of Minnesota Press.

Said, E. W. (1993). *Culture and imperialism.* New York: Knopf.

Sharma, A. (2006). Girl seeking suitable boy: Indian marriage dot com. Unpublished doctoral dissertation, University of Toronto, Canada.

Shukla, S. (2003). *India abroad: Diasporic cultures of postwar America and England.* Princeton, NJ: Princeton University Press.

Silver, D. (2000). Margins in the wires: Looking for race, gender, and sexuality in the Blacksburg Electronic Village. In B. E. Kolko, L. Nakamura & G. B. Rodman (Eds.), *Race in cyberspace* (pp. 133–150). London: Routledge.

Sudha, S. (1993). Compu-devata: Electronic bulletin boards and debates. *SAMAR (South Asian Magazine for Action and Reflection), 2*, 4–10.

Thapar, R. (2002). *Early India: From the origins to AD 1300.* Berkely: University of California Press.

Thompson, K. (2002). Border crossings and diasporic identities: Media use and leisure practices of an ethnic minority. *Qualitative Sociology, 25(3)*, 409–418.

Werbner, P. (1997). Introduction: The dialectics of cultural hybridity. In P. Werbner & T. Modood (Eds.), *Debating cultural hybridity: Multi-cultural identities and the politics of anti-racism* (pp. 1–28). London: Zed Books.

Chapter Nine

Anderson, B. (1999). *Imagined communities: Reflections on the origin and spread of nationalism.* New York: Verso.

Brown, J. M. (2003). *Nehru: A political life.* New Haven, CT: Yale University Press.

Central Intelligence Agency, Office of Public Affairs. (2005). Nepal. Retrieved February 21, 2005, from http://www.cia.gov/cia/publications/factbook/geos/np.html.

Chakravartty, P. (2001). The emigration of high-skilled Indian workers to the United States: Flexible citizenship and India's information economy. In W. A. Cornelius, T. J. Espenshade, & I. Salehyan (Eds.), *The international migration of the highly skilled: Demand, supply, and development consequences in sending and receiving countries* (pp. 325–349). La Jolla, CA: Center for Comparative Immigration Studies.

Das, A. N. (2002). The end of geography: Nationalism in the era of globalization. In R. Starr (Ed.), *Nations under siege: Globalization and nationalism in Asia* (pp. 31–62). New York: Palgrave.

Dutt, E. (n.d.). 28.4 million foreign-born residents in U.S. Retrieved February 21, 2005, from http://www.iacfpa.org/p_news/nit/iacpa-archieve/misc/24052002-census-1.html

Ganpati, P. & Basu, R. (2004). Pravasi Bharatiya Divas: Economy and the diaspora. Retrieved February 21, 2005, from http://specials.rediff.com/money/2004/jan/10sld1.htm.

Giddens, A. (2003). *Runaway world: How globalization is reshaping our lives.* New York: Routledge.

Hansen, T. B. (1999). *The saffron wave: Democracy and Hindu nationalism in modern India.* Princeton, NJ: Princeton University Press.

Khilnani, S. (1999). *The idea of India.* New Delhi: Penguin.

Lal, V. (2001). The politics of history on the Internet: Cyber-diasporic Hinduism and the North American Hindu diaspora. In M. R. Paranjpe (Ed.), *In diaspora: Theories, histories, texts* (pp. 179–221). New Delhi: Indialog.

Mathew, B. & Prashad, V. (2000). The protean forms of Yankee Hindutva. *Ethnic and Racial Studies, 23,* 516–534.

Mazumdar, S. (2003). The politics of religion and national origin: Rediscovering Hindu Indian identity in the United States. In V. Kaiwar & S. Mazumdar (Eds.), *Antinomies of modernity: Essays on race, Orient, nation* (pp. 223–260). Durham, NC: Duke University Press.

Nanda, M. (2003). *Prophets facing backward: Postmodern critiques of science and Hindu nationalism in India.* New Brunswick, NJ: Rutgers University Press.

Nehru, J. (1965). From address to youth, all-Bengal students conference, Calcutta, September 22. In D. Norman (Ed.), *Nehru, the first sixty years; presenting in his own words the development of the political thought of Jawaharlal*

Nehru and the background against which it evolved (Vol. 1) (pp. 161–167). London: Bodley Head.

Nehru, J. (1995/1946). *The discovery of India*. New Delhi: Jawaharlal Nehru Memorial Fund and Oxford University Press.

Neufeldt, R. (2003). The Hindu Mahasabha and Gandhi. In H. Coward (Ed.), *Indian critiques of Gandhi* (pp. 131–152). Albany: State University of New York Press.

Office of the Registrar General, Government of India. (2004). Profiles by main religion: Hindu. *Census of India 2001: Data on religion*, September 6. Retrieved February 20, 2005, from http://www.censusindia.net/religiondata/Summary%20Hindus.pdf.

Prakash, G. (1999). *Another reason: Science and the imagination of modern India*. Princeton, NJ: Princeton University Press.

Puri, B. (2001). Indian nationalism and the Hindu argument: Savarkar and Gandhi. In B. Puri (Ed.), *Mahatma Gandhi and his contemporaries* (pp. 154–176). Shimla: Indian Institute of Advanced Study.

Rajagopal, A. (2000). Hindu nationalism in the US: Changing configurations of political practice. *Ethnic and Racial Studies, 23(3)*, 467–496.

Saraswat, S. P. (n.d.). Mahabharat: An astronomical proof from the Bhagavat Puraan. Retrieved May 24, 2005, from http://www.hindunet.org/hindu_history/ancient/mahabharat/mahab_sarasvat.html.

Savarkar, V. D. (1967). Message on the eve of 59th birthday. In S. S Savarkar & G. M. Joshi (Eds.), *Historical statements—V. D. Savarkar* (pp. 1–2). Mumbai: Popular Prakashan.

Savarkar, V. D. (1969). *Hindutva: Who is a Hindu?* New Delhi: Bharatiya Sahitya Sadan.

Vanaik, A. (1997). *The furies of Indian communalism: Religion, modernity, and secularization*. New York: Verso.

Vartak, P. V. (n.d. a). *Astronomical dating of Ramayan*. Retrieved May 24, 2005, from http://www.hindunet.org/hindu_history/ancient/ramayan/rama_vartak.html.

Vartak, P. V. (n.d. b). The scientific dating of the Mahabharat war. Retrieved May 24, 2005, from http://www.hindunet.org/hindu_history/ancient/mahabharat/mahab_vartak.html.

Chapter Ten

Anderson, B. (1991). *Imagined communities: Reflections on the origin and spread of nationalism*. London: Verso.

Appadurai, A. (1996). *Modernity at large: Cultural dimensions of globalization*. Minneapolis: University of Minnesota Press.

Bastian, M. L. (1999). Nationalism in a virtual space: Immigrant Nigerians on the Internet. *West Africa Review, 1(1)*. Retrieved March 6, 2005, from http://www.westafricareview.com/vol1.1/1.1war.htm.

Da Rosa, V., Gold, G. L., & Lamy, P. (2000, April). *Ethnicity in a globalizing world: Borders, boundaries, and virtual communities.* Paper presented at the 40th Annual Meetings of the Northeastern Anthropological Association, York College (CUNY), New York.

Gajjala, R. (2004). *Cyber selves: Feminist ethnographies of South Asian women.* Walnut Creek, CA: AltaMira Press.

Granovetter, M. (1973). The strength of weak ties. *American Journal of Sociology, 78(6)*, 1360–1380.

Karamcheti, I. (1992). The shrinking Himalayas. *Diaspora, 2(2)*, 269–280.

Lak, D. (1999). Life and death in Orissa. BBC News, November 15 [transcript]. Retrieved March 6, 2005, from http://news.bbc.co.uk/hi/english/world/from_our_own_correspondent/newsid_520000/520523.stm.

Levitt, P. (2001). *The transnational villagers.* Berkeley: University of California Press.

Marx, L. (1999). Information technology in historical perspective. In D. A. Schön, B. Sanyal, & W. J. Mitchell (Eds.), *High technology and low-income communities: Prospects for the positive use of advanced information technology* (pp. 131–150). Cambridge, MA: MIT Press.

Mulgan, G. (1998). *Connexity: How to live in a connected world.* Boston: Harvard Business School Press.

Putnam, R. D. (2000). *Bowling alone: The collapse and revival of American community.* New York: Touchstone Books.

Rheingold, H. (2000). *The virtual community: Homesteading on the electronic frontier.* Reading, MA: Addison-Wesley.

Sterne, J. (1999). Thinking the Internet: Cultural studies versus the millennium. In S. Jones (Ed.), *Doing Internet research: Critical issues and methods for examining the Internet* (pp. 257–288). Thousand Oaks, CA: Sage.

Chapter Eleven

CBS News. (2004, August 1). Out of India. Retrieved March 10, 2006, from http://www.cbsnews.com/stories/2003/12/23/60minutes/main590004.shtml.

Chen, Z. (2004). Chinese minority and everyday racism in Canadian towns and small cities: An ethnic study of the case of Peterborough, Ontario, 1892–1951. *Canadian Ethnic Studies, 36(1)*, 71–102.

Cruz-Neira, C., Sandin D. J., & DeFanti, T. A. (1993). Surround-screen projection-based virtual reality: The design and implementation of the CAVE. *Proceedings of the 20th Annual Conference on Computer Graphics and Interactive Techniques* (pp. 135–142). Retrieved March 10, 2006, from http://portal.acm.org/results.cfm?coll=portal&dl=ACM&CFID=71126225&CFTOKEN=16997636.

Desai, A. V. (2005, November 15). Outsourcing your life—BPO work can be exciting, stimulating and enjoyable. *The Telegraph* (Calcutta, India). Retrieved

March 10, 2006, from http://www.telegraphindia.com/1051115/asp/opinion/story_5432742.asp.

Dudley, B. (2004, August 16). Microsoft's call-center business in India gets an American accent. *The Seattle Times.* Retrieved March 10, 2006, from http://seattletimes.nwsource.com/html/businesstechnology/2002006164_india english16. html.

Feenstra, R. C., & Hanson G. H. (1996). Productivity measurement and the impact of trade and technology on wages: Estimates for the U.S., 1972–1990. *Quarterly Journal of Economics, 114(3),* 907–940.

Fun loving brat! (2005, September 26). Indian English! It will be very helpful yaar! *India Times Blogs.* Retrieved March 10, 2006, from http://o3.indiatimes.com/feelgoodteen/archive/2005/09/26/266828.aspx.

Gentleman, A. (2005, October 30). Painful truth of the call centre cyber coolies. *The Observer.* Retrieved March 10, 2006, from http://observer.guardian.co.uk/international/story/0,6903,1604664,00.html.

Gil, A. G. & Vega, W. A. (1996). Two different worlds: Acculturation stress and adaptation among Cuban and Nicaraguan families. *Journal of Social and Personal Relationships, 13,* 435–456.

Goto, S. G., Gee, G. C., & Takeuchi, D. T. (2002). Strangers still? The experience of discrimination among Chinese Americans. *Journal of Community Psychology, 30(2),* 211–224.

Grossman, M. G., & Helpman, E. (2002). Outsourcing in a global economy. NBER Working Paper number 8278. Retrieved March 11, 2006, from http://www.nber.org/papers/w8728.

Hummels, D., Rapoport, D., & Yi, M. K. (1998). Vertical specialization and the changing nature of world trade. *Economic Policy Review, 4(2),* 79–99. Retrieved March 11, 2006, from http://www.newyorkfed.org/research/epr/98v04n2/9806humm.html.

Infocity. (n.d.). Retrieved March 10, 2006, from http://www.theinfocity.net/index.htm.

Joseph, M. (2002, October 23). Say hello to Sanjeep, Er, Sam. *Wired News.* Retrieved March 10, 2006, from http://www.wired.com/news/business/0,1367,55799,00.html.

Karpinski, E. C. (1996). Negotiating the self: Eva Hoffman's *Lost in Translation* and the question of immigrant autobiography. *Canadian Ethnic Studies, 28(1),* 172–201.

King, G. (1996). *Mapping reality: An exploration of cultural cartographies.* New York: St. Martin's Press.

Leslie, J. (2004, July 22). Georgia outsources jobs to India. Retrieved March 10, 2006, from http://www.11alive.com/specials/usworld/india/india_article.aspx?storyid=49513.

Lum, C. (1991). Communication and cultural insularity: The Chinese immigrant experience. *Critical Studies in Mass Communication, 8(1),* 91–101.

McPhate, M. (2005, November 17). Outsourcing outrage: Indian call-center workers suffer abuse. *San Francisco Chronicle*, p. C1.

Mitra, A., & Schwartz, R. L. (2001). From cyber space to cybernetic space: Rethinking the relationship between real and virtual spaces. *Journal of Computer Mediated Communication, 7(1).* Retrieved March 11, 2006, from http://jcmc.indiana.edu/vol7/issue1/mitra.html.

Mitra, A. (1999). Characteristics of the WWW text: Tracing discursive strategies. *Journal of Computer Mediated Communication, 5(1).* Retrieved March 11, 2006, from http://www.ascusc.org/jcmc/vol5/issue1/mitra.html.

Mukherjee, S. (2004, November 18). I'll never work at a call centre again! Retrieved March 10, 2006, from http://in.rediff.com/getahead/2004/nov/18ga-bpo.htm.

Negroponte, N. (1995). *Being digital.* New York: Vintage Books.

Nwagbara, U., Buehlmann, U., & Schuler, A. (2002). *The impact of globalization on North Carolina's furniture industries.* Raleigh, NC: North Carolina Department of Commerce, Policy, Research, and Strategic Planning Division and North Carolina State University. Retrieved January 15, 2004 from http://cmedis.commerce.state.nc.us/research/

Pathak, S. (2004, July 27). Calling that thrives on calls. *The Tribune* [Electronic Version]. Retrieved March 10, 2006, from http://www.tribuneindia.com/2004/20040727/jobs/main1.htm.

Pradhan, J. P., & Abraham, V. (2005). Social and cultural impact of outsourcing: Emerging issues from Indian call centers. *Harvard Asia Quarterly, 9(3).* Retrieved March 10, 2006, from http://www.asiaquarterly.com/content/view/155/5.

Shapiro, M. J. (1997). *Violent cartographies: Mapping cultures of war.* Minneapolis: University of Minnesota Press.

Sheffer, G. (2003). *Diaspora politics: At home abroad.* Cambridge: Cambridge University Press.

Spaans, J. A. (2000). GPS: The holy grail? *Journal of Navigation, 53*, 293–297.

Tye, L. (2001). *Home lands: Portraits of the new Jewish diaspora.* New York: Holt.

Wood, D. (1992). *The power of maps.* New York: Guilford Press.

Chapter Twelve

Adam, B. (1998). *Timescapes of modernity: The environment and invisible hands.* London: Routledge.

Adam, B. (2002). The gendered time politics of globalization: Of shadowlands and elusive justice. *Feminist Review, 70*, 2–29.

Appadurai, A. (1996*). Modernity at large: Cultural dimensions of globalization.* Minneapolis: University of Minnesota Press.

Appadurai, A. (2000). Grassroots globalization and the research imagination. *Public Culture, 12(1)*, 1–19.

Bain, P. & Taylor, P. (2002). Consolidation, "cowboys" and the developing employment relationship in British, Dutch and US call centres. In U. Holtgrewe (Ed.), *Re-organizing service work: Call centers in Germany and Britain* (pp. 42–62). Burlington, VT: Ashgate.

Barnes, A. (2004, April). The construction of control: The impact of the physical environment on the development of resistance and accommodation within call centers. Paper presented at the 22nd International Labour Process Conference, University of Amsterdam.

Bergeron, S. (2001). Political economy discourses of globalization and feminist politics. *Signs: Journal of Women in Culture and Society, 26(44)*, 983–1006.

Bhabha, H. K. (1994). *The location of culture*. London: Routledge.

Bradley, H. (2000). *Myths at work*. Cambridge: Polity.

Brah, A. (2002). Global mobilities, local predicaments: Globalization and the critical imagination. *Feminist Review, 70,* 30–45.

Chhachhi, A. (1999). Gender, flexibility, skill and industrial restructuring: The electronics industry in India. *Gender, Technology and Development, 3(3),* 329–360.

Datta, R. C. (2004, April). Worker and work: A case study of an international call centre in India. Paper presented at the 22nd International Labour Process Conference, University of Amsterdam.

Freeman, C. (2000). *High tech and high heels in the global economy: Women, work and pink collar identities in the Caribbean*. Durham, NC: Duke University Press.

Freeman, C. (2001). Is local: Global as feminine: Masculine? Rethinking the gender of globalization. *Signs: Journal of Women in Culture and Society, 26(4),* 1007–1038.

Gibson-Graham, J. K. (1996). *The end of capitalism (as we know it)*. Cambridge, MA: Blackwell.

Grewal, I. & Kaplan, C. (1994). Introduction: Transnational feminist practices and questions of postmodernity. In I. Grewal & C. Kaplan (Eds.), *Scattered hegemonies: Postmodernity and transnational feminist practices* (pp. 1–36). Minneapolis: University of Minnesota Press.

Hall, S. (1996). The West and the rest: Discourse and power. In S. Hall, D. Held, D. Hubert, & K. Thompson (Eds.), *Modernity: An introduction to modern societies* (pp.184–228). Oxford: Blackwell.

Harvey, D. (1989). *The condition of postmodernity: An enquiry into the origins of cultural change*. Oxford: Blackwell.

Katz, C. (2001). On the grounds of globalization: A topography for feminist political engagement. *Signs: Journal of Women in Culture and Society, 26(4),* 1213–1234.

Kaushalesh, L. (2001). Institutional environment and development of information and communication technology in India. *Information Society, 17*, 105–117.

Kondo, D. K. (1990). *Crafting selves: Power, gender, and discourses of identity in a Japanese workplace*. Chicago: University of Chicago Press.

Larner, W. (2002). Calling capital: Call centre strategies in New Brunswick and New Zealand. *Global Networks, 2(2)*, 133–153.

Leidner, L. (1999). Emotion labour in service work. *Annals of the American Academy of Political and Social Science, 56(1)*, 81–95.

Mitter, S. (2000, June 24). Telework, teletrade and sustainable development: The Indian experience in a global context. *Economic and Political Weekly, 35(26).*

Nanda, M. (2000). Post-Fordist technology and the changing patterns of women's employment in the third world. *Gender, Technology and Development, 4(1)*, 25–56.

NASSCOM (National Association of Software and Service Companies). (2001). *NASSCOM's handbook: IT enabled service background and reference resource.* Retrieved March 25, 2006, from http://www.nasscom.org.

Nayar, B. R. (2003). Globalization and India's national autonomy. *Commonwealth and Comparative Politics, 41(2)*, 1–34.

Ong, A. (1991). The gender and labor politics of postmodernity. *Annual Review of Anthropology, 20,* 279–309.

Phillipson, R. (2001). Global English and local language policies: What Denmark needs. *Language Problems and Language Planning, 25(1)*, 1–24.

Sarker, S. & Niyogi De, E. (2002). *Trans-status subjects: Gender and the globalization of South and South East Asia.* Durham, NC: Duke University Press.

Sassen, S. (2000). Spatialities and temporalities of the global: Elements for a theorization. *Public Culture, 12(1)*, 215–232.

Sassen, S. (2001). Cracked casings notes towards an analytics for studying transnational processes. In L. Pries (Ed.), *Transnational social spaces, international migration and transnational companies in the early 21st century* (pp. 187–208). New York: Routledge.

Sassen, S. (2002, November 29). The formation of new political subjects under globalization. Public lecture at the Institute for Women's Studies and Gender Studies, University of Toronto.

Scott, J. C. (1990). *Domination and the arts of resistance: Hidden transcripts.* New Haven: Yale University Press.

Swales, J. M. (1997). English as Tyrannosaurus rex. *World English, 16(3)*, 373–382.

Taylor, P. & Bain, P. (2004, April). India calling to the far away towns: The call centre labour process and globalization. Paper presented at the 22nd International Labour Process Conference, University of Amsterdam.

Wright, M. (2001). Asian spies, American motors, and speculations on the space-time of value. *Environment and Planning A, 33(12)*, 2175–2188.

Chapter Thirteen

Ahmad, T. (2001). The state of Internet economy in India. In O. Manzar, M. Rao, and T. Ahmad (Eds.), *The Internet economy of India* (pp. 3–9). New Delhi: Inomy Media Pvt Ltd.

Appiah, K. A. (1996). Is the post in postmodernism the post in postcolonial? In P. Mongia (Ed.), *Contemporary postcolonial theory: A reader* (pp. 55–71). Oxford: Oxford University Press.

Bhabha, H. K. (1994). *The location of culture.* New York: Routledge.

Butler, J. (1997). *The psychic life of power: Theories in subjection.* Stanford, CA: Stanford University Press.

Click, R. L. & Duening, T. N. (2005). *Business process outsourcing: The competitive advantage.* Hoboken, NJ: John Wiley.

Foster, R. J. (1991). Making national cultures in the global ecumene. *Annual Review of Anthropology, 20,* 235–260.

Friedman, T. (2005). *The world is flat: A brief history of the twenty-first century.* New York: Farrar, Straus & Giroux.

Friedman, T. (2006, May 21). Whole world grows into one big office. *The News Tribune, Insight,* p. 3.

McMillin, D. C. (2001). Localizing the global: Television and hybrid programming in India. *International Journal of Cultural Studies, 4(1),* 45–68.

McMillin, D. C. (2003). Television, gender, and labor in the global city. *Journal of Communication, 53(3),* 496–511.

Miller, R. R. (2001). Leapfrogging? India's information technology industry and the Internet. Discussion Paper number 42. Washington, DC: International Finance Corporation.

NASSCOM (National Association of Software and Service Companies). (2003). *The IT industry in India.* Strategic Review 2003. New Delhi: NASSCOM.

Paley, A. (2006, May 17). Latest overseas outsourcing? Tutoring. *The News Tribune,* p. A8.

Said, E. (1979). *Orientalism.* New York: Vintage Books.

Schiller, H. I. (1991). Not yet the post-imperialist era. *Critical Studies in Mass Communication, 8,* 13–28.

Schwarz, H. (2000). Mission impossible: Introducing postcolonial studies in the US academy. In H. Schwarz & S. Ray (Eds.), *A companion to postcolonial studies* (pp. 1–20). Malden, MA: Blackwell Publishers.

Sharp, D. (2003). *Call center operation: Design, operation, and maintenance.* New York: Digital Press.

Singh, P. & Pandey, A. (2005, February 12). Women in call centers. *Economic and Political Weekly, 40(7).*

Strinati, D. (1995). *An introduction to theories of popular culture.* London & New York: Routledge.

Verma, S. & Sharma, D. (2003). Cultural continuity amid social change: Adolescents' use of free time in India. In S. Verma & R. Larson (Eds.), *Examining adolescent leisure time across cultures: Developmental opportunities and risks* (pp. 37–51). San Francisco: Wiley.

Wallerstein, I. (1990a). Culture and the ideological background. In M. Featherstone (Ed.), *Global culture: Nationalism, globalization, and modernity* (pp. 31–55). New York: Sage.

Wallerstein, I. (1990b). Culture is the world-system: A reply to Boyne. In M. Featherstone (Ed.), *Global culture: Nationalism, globalization, and modernity* (pp. 63–65). New York: Sage.

Conclusion

Berry, C., Martin, F., & Yue, A. (Eds.). (2003). *Mobile cultures: New media in queer Asia.* Durham, NC: Duke University Press.

Castranova, E. (2006). *Synthetic worlds: the business and culture of online games.* Chicago: University of Chicago Press.

Dibbell, J. (2006). *Play money: Or, how I quit my day job and made millions trading virtual loot.* New York: Basic Books.

Dirlik, A. (2006). Introduction: Our ways of knowing—and what to do about them. In A. Dirlik (Ed.), *Pedagogies of the global* (pp. 3–18). Boulder, CO: Paradigm Publishers.

Gajjala, R. (2006). Editorial: Consuming/producing/inhabiting South-Asian digital diasporas. *New Media and Society, 8(2)*, 179–185.

Poster, M. (1995). *The second media age.* Cambridge: Polity Press.

Rai, A. (1995). India on-line: Electronic bulletin boards and the construction of diasporic Hindu identity. *Diaspora, 4(1),* 31–58.

Scheidt, L. A. (2006). Adolescent diary weblogs and the unseen audience. In D. Buckingham & R. Willett (Eds.), *Digital generations: Children, young people and new media.* London: Lawrence Erlbaum.

Taylor, T. L. (2006). *Play between worlds: Exploring online game culture.* Cambridge: MIT Press.

CONTRIBUTORS

Vinita Agarwal (M.A., University of Illinois, Chicago) is a Ph.D. candidate in the Department of Communication at Purdue University. Her research interests center at the intersection of health care quality, technology and media, particularly in the context of organizational and social change. Her work has been published in the *Western Journal of Communication* and in an edited collection on linguistic and online cultural issues.

Patrice M. Buzzanell (Ph.D., Purdue University) is Professor in the Department of Communication at Purdue University, where she specializes in gendered workplace issues, particularly those associated with career, leadership, and work-life concerns. She has published in a number of journals and edited collections including *Communication Monographs, Human Communication Research, Communication Theory, Handbook of Applied Communication Research, and Human Relations.* She is the incoming President of the International Communication Association and the Council of Communication Associations

Rohit Chopra (Ph.D., Emory University) is Assistant Professor of Media Studies at Babson College, Massachusetts. He is the author of *Mediated Visions: Technology and Identity Politics in India from Colonialism to the Internet Age*, forthcoming from Cambria Press (2008/2009). He has published articles in several journals, including the *Economic and Political Weekly, New Media and Society, First Monday,* and *Cultural Studies.*

Jillana Enteen (Ph.D., Rutgers University) is a Senior Lecturer for the Gender Studies Program at Northwestern University. She writes about Internet use by overlooked populations and how this changes English language conceptions of gender, sexuality, nation, and ethnicity. Her essays have been published in journals such as *New Media and Society, Science Fiction Studies,* and *The Journal of Lesbian Studies.* Her book, *Virtual English: Language, Internet*

Use and Digital Creolization, is forthcoming (2008) on Routledge Press.

Radhika Gajjala (Ph.D., University of Pittsburgh) is Associate Professor and Graduate Coordinator in the School of Communication studies at Bowling Green State University, Ohio. Her book *Cyberselves: Feminist Ethnographies of South Asian Women* (2004) was published by Altamira Press. She is currently working on a single-authored book on *Technocultural Agency: Production of Identity at the Interface* (under contract with Lexington) and on an edited volume *Digital Embodiment in 3D Worlds*.

Venkataramana Gajjala (Ph.D., University of Pittsburgh) is Associate Professor in the School of Business at Tiffin University, Ohio. He has work experience in banking and in the corporate sector in both India and the U.S. He is the author of journal articles in the field.

Linda Leung is a Senior Lecturer at the Institute for Interactive Media & Learning, and member of the Centre for Human-Centred Technology Design at the University of Technology, Sydney. She is author of *Virtual Ethnicity: Race, Resistance & the World Wide Web* (2005).

Divya C. McMillin is author of *International Media Studies* (2007), a postcolonial critique of research in international communication. Her articles in the *Journal of Communication, International Journal of Cultural Studies*, and *Continuum: Journal of Media and Cultural Studies*, to name a few, address media globalization, neocolonialism, national identity, and television reception.

Kiran Mirchandani (Ph.D., McGill University) is an Associate Professor at the Ontario Institute for Studies in Education of the University of Toronto. She has published on home-based work, telework, contingent work, entrepreneurship, transnational service work, and self-employment. Her articles have appeared in journals such as *Gender & Society*, *Global Networks*, *Qualitative Methods* and *Gender, Work and Organization*. She teaches in the

Adult Education and Community Development Program (workplace learning and change focus), and offers courses on gendered and racialized processes in the workplace; critical perspectives on organizational development and learning; and technology, globalization, and economic restructuring.

Ananda Mitra (Ph.D., University of Illinois at Urbana-Champaign) is Professor of Communication at Wake Forest University. His research focuses on the cultural and social implications of living in cybernetic space. He teaches courses on new technologies, popular culture, research methodology, and issues about South Asia.

Anustup Nayak is a Partner with iDiscoveri Center for Education and Enterprise in India, a social enterprise working to bring innovative teaching and learning solutions to schools in India and other developing countries. Anustup leads iDiscoveri's practice in school reform and education leadership. Anustup holds master's degrees from Harvard Graduate School of Education and Georgia Tech School of Public Policy.

Amit S. Rai (Ph.D., Stanford University) teaches in the Program in The History of Media Ecologies in the English Department at Florida State University. His first book, *Rule of Sympathy: Sentiment, Race and Power, 1760-1860* (St. Martins-Palgrave, 2002), is a political history of sentiment and humanitarianism in different forms of colonial power (West Indian slavery and missionizing evangelicalism in India). His most recent articles include a new narratology of gender, sexuality, and power in contemporary diasporic Asian film narratives (*South Asian Popular Culture*); a questioning of the figure of the Muslim in Hindi films (*Harvard Asia Quarterly*); analyses of monstrosity and terrorism (*Cultural Studies and in Social Text*); and occasional reviews and essays for Art India, Humanscape (Mumbai, India) and SAMAR (New York). His forthcoming study of media in India entitled *Untimely Bollywood: Globalization and India's New Media Assemblage* is due out in the spring of 2008 from Duke University Press.

Natalia Rybas is a doctoral candidate at the School of Communication Studies, Bowling Green State University (Ohio). Her research interests concentrate around cyberculture, technology-mediated communication, critical studies of technology, and cyberethnography. Currently, she is working on her dissertation on technocultural practices in social network systems.

Archana Sharma holds a Ph.D. in Sociology of Education from the Ontario Institute for Studies in Education, at the University of Toronto, Canada. Her areas of expertise are cultural studies, (post)colonial studies, immigration, and media studies, ethnic studies, and antiracist education. Her work examines transnational communities such as diasporic Indians in Canada and the U.S. and the role the Internet and/or technology has played in sustaining, negotiating, and reformulating ethnic identities. Specifically she writes on how a desire for community and belonging complement and/or resist hegemonic notions of class, caste, religion, race, nation, gender, and sexuality. Her research has been mainly concerned with cultural approaches and uses of technology.

P. Thirumal is Associate Professor of Communication, University of Hyderabad. His research publications and interests are in the areas of Media Historiograhy, Dalit Identity and Media Use, Discursive Production of Region and Other Cultural Geographies

Nabeel Zuberi teaches in the Department of Film, Television and Media Studies at the University of Auckland. His publications include *Sounds English: Transnational Popular Music* (University of Illinois Press, 2001) and *Media Studies in Aotearoa/New Zealand* (with Luke Goode, Pearson, 2004). He is currently working on *Understanding Popular Music*, a book for Sage Publications.

INDEX

Digital Formations

General Editor: *Steve Jones*

Digital Formations is an essential source for critical, high-quality books on digital technologies and modern life. Volumes in the series break new ground by emphasizing multiple methodological and theoretical approaches to deeply probe the formation and reformation of lived experience as it is refracted through digital interaction. Digital Formations pushes forward our understanding of the intersections—and corresponding implications—between the digital technologies and everyday life. The series emphasizes critical studies in the context of emergent and existing digital technologies.

Other recent titles include:

Leslie Shade
Gender and Community in the Social Construction of the Internet

John T. Waisanen
Thinking Geometrically

Mia Consalvo & Susanna Paasonen
Women and Everyday Uses of the Internet

Dennis Waskul
Self-Games and Body-Play

David Myers
The Nature of Computer Games

Robert Hassan
The Chronoscopic Society

M. Johns, S. Chen, & G. Hall
Online Social Research

C. Kaha Waite
Mediation and the Communication Matrix

Jenny Sunden
Material Virtualities

Helen Nissenbaum & Monroe Price
Academy and the Internet

To order other books in this series please contact our Customer Service Department:

(800) 770-LANG (within the US)
(212) 647-7706 (outside the US)
(212) 647-7707 FAX

To find out more about the series or browse a full list of titles, please visit our website:

WWW.PETERLANG.COM